AF256375

SYNTHETIC BILOGY

A Lab Manual

Second Edition

SYNTHETIC BIOLOGY

A Lab Manual

Second Edition

Anthony C. Forster
Uppsala University, Sweden

Letian Bao
Uppsala University, Sweden

Josefine Liljeruhm
Uppsala University alumna, Sweden

NEW JERSEY · LONDON · SINGAPORE · BEIJING · SHANGHAI · HONG KONG · TAIPEI · CHENNAI

Published by

World Scientific Publishing Europe Ltd.

57 Shelton Street, Covent Garden, London WC2H 9HE

Head office: 5 Toh Tuck Link, Singapore 596224

USA office: 27 Warren Street, Suite 401-402, Hackensack, NJ 07601

British Library Cataloguing-in-Publication Data
A catalogue record for this book is available from the British Library.

SYNTHETIC BIOLOGY
A Lab Manual (Second Edition)

ISBN 978-1-80061-535-9 (hardcover)
ISBN 978-1-80061-672-1 (paperback)
ISBN 978-1-80061-673-8 (ebook for institutions)
ISBN 978-1-80061-674-5 (ebook for individuals)

For any available supplementary material, please visit
https://www.worldscientific.com/worldscibooks/10.1142/Q0493#t=suppl

About the Authors

Anthony C. Forster (M.D., Harvard; B.Sc.Hons., Ph.D. in Biochemistry, Adelaide) is a professor researching synthetic biology at the Department of Cell and Molecular Biology, Uppsala University, Sweden. He discovered the hammerhead catalytic RNA structure, authored patents that founded two biotech companies, edited synthetic biology volumes of *Methods* and *Biotechnology J.*, and created the synthetic biology lab course detailed in this manual.

Letian Bao (B.Biotech., Jilin, China; M.Biol., Uppsala) is a Ph.D. student in Prof. Forster's lab at the Department of Cell and Molecular Biology, Uppsala University, Sweden. He engineered chromoproteins to overcome their limitations and improve the lab course.

Josefine Liljeruhm (M.Sc. in Molecular Biotechnology and Ph.D. in Cell and Molecular Biology, Uppsala). As a Ph.D. student in Prof. Forster's lab, she set up and taught the lab course detailed in this manual and characterized the chromoproteins.

Preface to the 2nd edition

A decade has passed since the publication of the first edition (Liljeruhm *et al.*, 2014) of the first lab manual in synthetic biology. Thus, although the manual was well received (Hsueh and Helmke, 2014) and continues to serve students and starting researchers well, important updates were needed. We have identified major toxicity problems with chromoproteins and reporters used in the lab course, and overcame the toxicity challenges through engineering. We made these second-generation versions available as a kit from Addgene. Regarding DNA assembly, Type IIS cloning schemes became increasingly favored over BioBricks, but the latter method is simpler and we showed BioBricks to be scalable (front cover). Commercial multigene gene synthesis and assembly in yeast became so routine that even refactored versions of entire yeast chromosomes were synthesized. CRISPR targeting, invented just before the 1st edition (Jinek *et al.*, 2012), dramatically simplified editing of eukaryotic genomes. The breakthrough was further improved by engineering and is now available in kits for the "biohacker" public. This is just the sort of Nobel Prize-winning SynBio revolution anticipated by George Church in his Foreword to the 1st edition. Our goal was not to duplicate every new and old biomolecular recipe: that would require an encyclopedia, not a handbook. Rather, we focused on key SynBio methods needed to get you up and running quickly in the lab — and succeeding!

Anthony C. Forster

Foreword to the 1st edition

This Foreword began when Tony Forster and I met a decade ago asking each other "What is Life?" (and "minimal life?") from the genetic, bioinformatics, and biochemical perspective. We soon began co-authoring a vision statement (eventually two articles) to accompany an experimental paper in the journal *Nature* which described the first assembly of genes (and a long operon) from oligonucleotides synthesized on and extracted from "glass chips" (the pattern and order of A, C, G, T photochemically programmed). More recently, I have participated (via Skype) in Tony's class on synthetic biology in Uppsala using the book *Regenesis* as a text (which I co-authored with Ed Regis). The question-and-answer sessions were very lively and thought-provoking, but there was no doubt that the field of Synthetic Biology needed at that time "A Lab Manual," such as the one that you now hold in your hands. The excitement of SynBio extends beyond even high school, college courses and competitions like the international Genetically Engineered Machine (iGEM), ramifying into the broad and unpredictable world of do-it-yourself biology. What garage holds the next Steve Jobs and Woz? What Lab Manual will be at their fingertips?

In addition to the definitions of synthetic biology in the Preface and Introduction, I'd add these: synthetic biology is to early recombinant DNA as a genome is to a gene. The formation of SynBio as a discipline finally earns the label "genetic engineering." SynBio soars beyond making old biology "easier," and moves onward to also more accurate and effective goals. Biological engineering is poised to outdo all previous engineering fields because of three advantages: (1) SynBio has the ability to miniaturize in 3 and 4 dimensions via multiplexing and molecular "libraries", well exceeding the capabilities of 2-dimensional microfabrication of electrical engineers, or the multidimensional, yet coarse-grained, constructions of mechanical and civil engineers. (2) SynBio has inherited billions of years of evolutionary innovation and testing — an unparalleled list of parts, systems and applications. (3) SynBio can combine (1) and (2) to produce accelerated evolution, like the MAGE (multiplex automated genome engineering) section of this book or "PACE" (phage-assisted continuous evolution). Clever selections and screens allow us to test not one prototype at a time, but billions. Moreover, we can use SynBio to evolve SynBio.

That auto-catalytic statement above brings us to the elephant (or mammoth) in the room — "Exponential Technologies." The rate of change of biotechnologies (about 8-fold per year) has exceeded even the super rate of Moore's Law for electronics (about 1.5-fold per year). Keep in mind that these are not "laws" of physics, but trend lines and can change slope dramatically up or down. For example, DNA sequencing was going at about.1.5-fold per year from its start in 1968 until 2005, when "next-generation-sequencing" began.

Safety. Such exponential change, especially in the context of lowered costs and greater accessibility, raises questions not just about conventional lab safety (as in Chapter 4 of this book), but potential for catastrophic scale bioerror or bioterror. As exhorted to the synthetic biologist Peter Parker by his Uncle Ben (quoting Voltaire), "with great power comes great responsibility." Democratization of SynBio and open publishing of promethean protocols mean that the vast majority of us who are virtuous and careful must be vigilant and creatively anticipate unintended consequences. It is not sufficient to say that we don't see any problem with our own work today.

The Future. This is the part of the Foreword reaching forward. Forewarned is forearmed. This is intentionally very "meta", meaning self-referential. The book *Regenesis* was written into DNA and 70 billion copies made (easily). You may want to encode *Synthetic Biology: A Lab Manual* into DNA, or annotate your next genetic construct with your lab notes encoded into the genome. Synthetic biology is branching out into realms far beyond single microbial cells — to synthetic communities of such cells, synthetic ecosystems, synthetic multicellular systems, synthetic developmental biology, gene therapy, resurrecting extinct species, perhaps even changing our own species. Synthetic biology is not limited to cells or even to carbon-based chemistry (e.g. the intricate inorganic patterns of diatoms, bones, sea shells, magneto-tactic particles in bacteria, silicon-based lenses and fiber optics in hexactinellid sponge). Biology inspires us to harness self-assembly, to turn industrial processes previously limited to enormous temperatures and toxic wastes to new "green chemistry" at 25°C. Protein/RNA machines and DNA origami represent, by far, the best nanotechnologies for rapidly designing and building atomically precise 3-dimensional objects, including mixtures of biopolymers and inorganic nanomaterials.

As electronics transitions from 2-dimensional to 3-dimensional space and ventures down from "bulk" (15 nm) photolithography and dopants, attention is turning to molecular computing. Biomolecular systems are a million times denser and lower energy per digital operation. The central question is transitioning from "What can synthetic biology do?" to "Is there anything that synthetic biology will not impact?"

George Church
Harvard Medical School

Preface to the 1st edition

Synthetic biology is an emerging multidisciplinary field with the potential to revolutionize our approach to global problems ranging from drug discovery to energy production. So what exactly is "SynBio," why is it so powerful, and why is it exciting students in ways never seen before? Perhaps the best answer is: synthetic biology is cutting-edge methods that make biology much easier to engineer. Yet there was no lab manual compiling these methods or detailing a synthetic biology lab course. Our goal was to fill this remarkable gap.

If you are one of the following: a teacher wanting to set up a student-driven synthetic biology lab course, a student taking such a lab course, a high school or tertiary institution student looking to participate in the iGEM competition, a new researcher in synthetic biology, or a non-specialist curious to know more about SynBio, iGEM and "BioBricks™," this book is for you. The protocols are straightforward enough to be used by any science students, be they biologists, chemists, engineers or even undifferentiated high school students.

Inspiration for writing this manual came from the grassroots level: students at Sweden's Uppsala University. They pressed the faculty to host five successive iGEM teams and to start a lab course, came up with the key project ideas, and thoroughly enjoyed themselves! The subject matter is also close to our hearts: ACF is a synthetic biology professor who created this lab course, JL is a Ph.D. student who set up and taught this lab, and

EG is a Ph.D. student who tutored Uppsala's last three iGEM teams. So we finish with a warning before opening: SynBio is highly addictive!

Josefine Liljeruhm
Erik Gullberg
Anthony C. Forster

Acknowledgments

We are very grateful for the groundswell of enthusiasm for synthetic biology at Uppsala University that ultimately made these book editions possible. In particular, we would like to thank Profs. Måns Ehrenberg and Leif Kirsebom for recruiting and mentoring ACF, Uppsala University undergraduate students for running stimulating iGEM projects; Prof. Peter Lindblad and Dr. Thorsten Heidorn for jointly hosting the first two of these iGEM teams; Prof. Anders Virtanen and Dr. Margareta Krabbe for hosting the other iGEM teams with ACF and helping ACF to launch a lab course in synthetic biology; Mirthe Hoekzema for suggesting using chromoproteins in the course; Anne-Maj Gustafsson and Peter Lillhager for managing the teaching labs; several doctoral students (particularly Jinfan Wang) and Dr. Victoriia Karpenko for teaching assistance jointly with JL or LB and for advice; Erik Lundin for summarizing iGEM protocols; Daniel Camsund and Elias Englund for advice on Gibson assembly; the course students cited for experimental results in some of the figures, the photographers cited in the figures; Dr. David Fange for creating a tutorial on computational modeling; and Prof. Dan Andersson for mentoring Erik Gullberg. Erik, now Dr. Wistrand-Yuen, was a major co-author of our first edition.

Beyond Uppsala, we thank Prof. emer. Anders Liljas and Sook-Cheng Lim for publishing advice and Prof. George Church for critiquing the first edition and writing an inspiring Foreword.

Contents

7. Advanced Methods   121

8. The international Genetically Engineered Machine (iGEM) Competition   135

9. Appendices   141

Abbreviations and Acronyms

3A	3 antibiotic
°C	degrees Celsius
A	adenine
amp	ampicillin
ASD	anti-Shine Dalgarno sequence
ATP	adenosine triphosphate
b	base
BB	BioBrick™
bp	base pair
BSA	bovine serum albumin
BSL	biosafety level
C	cytosine
Cas	CRISPR-associated
CDS	coding sequence
chloramp	chloramphenicol
CRISPR	clustered regularly interspaced short palindromic repeats
dATP	deoxyadenosine triphosphate
dCTP	deoxycytidine triphosphate
ddH$_2$O	doubly distilled water
dGTP	deoxyguanosine triphosphate
dH$_2$O	deionized water
DMSO	dimethyl sulfoxide

DNA	deoxyribonucleic acid
dNTP	deoxynucleoside triphosphate
dTTP	deoxythymidine triphosphate
E. coli	*Escherichia coli*
EDTA	ethylenediaminetetraacetic acid
FACS	fluorescence-activated cell sorting
fMet	formylmethionine
FRT	flippase recognition target
FW	formula weight
G	guanine
GFP	green fluorescent protein
GMO	genetically modified organism
iGEM	international Genetically Engineered Machine
IPTG	isopropyl ß-D-1-thiogalactopyranoside
kan	kanamycin
L, l	liter
LB	lysogeny broth
M	molar
m	milli (0.001x) or mass
MAGE	multiplex automated genome engineering
mRNA	messenger RNA
MSDS	material safety data sheet
MW	molecular weight
N	nucleotide
n	nano (0.000 000 001x)
NAD	nicotinamide adenine dinucleotide
nt(s)	nucleotide(s)
OD	optical density
OE-PCR	overlap extension PCR
oligo(s)	oligodeoxyribonucleotide(s)
ORF	open reading frame
PACE	phage-assisted continuous evolution
PAM	protospacer adjacent motif
PBS	phosphate buffered saline
PCR	polymerase chain reaction
PEG	polyethylene glycol

PNK	polynucleotide kinase
pol	polymerase
RBS	ribosome binding site
RFP	red fluorescent protein
RNA	ribonucleic acid
rRNA	ribosomal RNA
S	Svedberg unit
SD	Shine–Dalgarno sequence
SDS	sodium dodecyl sulphate
SOB	super optimal broth
sRNA	small RNA
SynBio	synthetic biology
T	thymine
TAE	tris(hydroxymethyl) aminomethane acetate ethylenediaminetetraacetate
TATA box	core promoter sequence 5'-TATAAA
TBE	tris(hydroxymethyl) aminomethane borate ethylenediaminetetraacetate
T_m	melting temperature
Tris, Trizma^R	tris(hydroxymethyl)aminomethane
tRNA	transfer RNA
TSS	transcription start site
U	uracil or university
UV	ultraviolet
V, v	volume
w	weight
x	times or multiplied by
μ	micro (0.000001x)

1

Introduction

What is Synthetic Biology, Exactly?

You may know it when you see it, but how would you define it? Synthetic biology is undoubtedly a new, rapidly growing field that is captivating students and researchers alike. Yet, like life itself, synthetic biology is notoriously hard to define. Though the term "synthetic biology" was coined a century ago, its use only came into vogue two decades ago. This "renaissance" cannot be attributed to any single breakthrough or publication, so why did it occur? How does synthetic biology differ from, for example, the older field of biotechnology that encompasses DNA cloning, polymerase chain reaction (PCR), monoclonal antibodies and protein overexpression? Engineers may emphasize **the development of foundational technologies that make the design and construction of engineered biological systems easier**" (Endy, 2005), such as BioBricks™ (Knight, 2003). Alternatively, biologists and chemists wishing to encompass both in vitro and in vivo projects may describe synthetic biology as "**the complex**

manipulation of replicating systems" (Forster and Church, 2007). Still, others define synthetic biology in terms of applications where, for example, **bioenergy, biomaterials and biosensors** are synthetic biology, while antibodies and induced pluripotent stem cells are not. The use of engineering principles for biological applications is not new, as evidenced by the long-term success of biotechnology and genetically modified organisms (GMOs) in agriculture. But what is definitely new versus classical recombinant DNA and PCR is that synthetic biology is much easier and more creative. The parts and techniques are more standardized and cheaper, allowing faster, more modular use with more predictable outcomes based upon more precise measurements of activities. Computer-aided design, analysis and modeling have further hastened progress. These next-generational technologies, together with inexpensive, rapid, commercial oligodeoxynucleotide/gene syntheses and high-throughput sequencing, tagged libraries, and CRISPR editing, have empowered biology and engineering students and scientists like never before.

The iGEM Outbreak

iGEM is an acronym for **international Genetically Engineered Machine** and is a worldwide annual competition in synthetic biology for students from secondary and tertiary institutions (Fig. 1). The Massachusetts Institute of Technology (MIT) organized the first competition in 2004 between teams from five universities in the U.S.A. Projects are student-driven and lab work is mostly done during the summer when students and labs are free from classes. iGEM has proven to be one of the most motivational educational methods ever devised, with the competition growing every year to now encompass 400 teams, including 50 teams in a high school division. It is also impacting industry, having been responsible for 200 startups.

In parallel with this infectious, grassroots movement, career opportunities in synthetic biology in industry and academia are ballooning, mandating more defined education in synthetic biology than achievable just through iGEM. Thus, student enthusiasts are teaming up with university administrators to demand the creation of formal courses in synthetic biology, with our full lab course beginning in 2013 as a result. Formal courses

Fig. 1   iGEM teams competing at MIT in 2006. Annual iGEM competitions have since expanded to encompass thousands of students worldwide. (Photograph by Randy Rettberg; taken from Wikipedia with permission.)

complement and differ from iGEM by providing a more rounded education, requiring individual responsibility for knowledge and lab skills, and awarding credit on an individual basis.

A Synthetic Biology Lab Manual

Even the most highly motivated synthetic biology students in general, and iGEM students in particular, can be pretty raw in their knowledge and require considerable guidance with lab work. And teachers with no prior experience in synthetic biology can be pressed with the daunting task of setting up an entire lab course from scratch. What was sorely needed was a lab manual. Here, we address this with three goals:

1. To provide **teachers and lab managers** with all the information they need to set up and run a synthetic biology lab course that spans 2 to 5 weeks or more of full-time work;
2. To provide **high school students and tertiary institution students and researchers** with protocols for lab work; and
3. To provide **iGEM students** with practical information on setting up and running their own summer project in a host lab.

The underlying philosophy of our lab course design is similar to that of iGEM: to foster learning through student-driven, creative research with cutting-edge methods in small teams. Teams encourage learning through discussions and teamwork, not to mention being a practical way of economizing use of reagents and equipment. **Students will not only learn synthetic biology technology such as BioBrick™ cloning, but also have the opportunity to create their own projects with varying difficulties.** Where the course philosophy differs from iGEM is in requiring each individual student to submit their own lab book for assessment and also to take a final exam. Without this individual responsibility, some team members will rely too heavily on other team members or become too specialized, thus failing to learn the important principles and failing to keep up with all aspects of the project. For example, an iGEM computational modeler may not understand operon function.

Just as iGEM teams struggle each year to pick a good project, so did we agonize in selecting a project for the lab course. Key considerations are outlined here as they may be helpful to future iGEM teams and course planners alike.

1. Expensive new equipment should not be required due to budget limitations.
2. In vivo replicating systems are generally easier to engineer in a novel way than in vitro ones. And most synthetic biology is in vivo.
3. *Escherichia coli* (*E. coli*) is the best model organism in biology. It is the best characterized, among the simplest and fastest growing, safe, and it is compatible with, by far, the most BioBrick™ DNA parts (already numbering in the thousands).
4. The use of chromoproteins as easy readouts of gene expression is encouraged. Chemical substrates are not required, neither is the ultra-violet (UV) light that is needed for optimal detection of the related fluorescent proteins, such as green fluorescent protein (GFP). Furthermore, the colors can be changed readily by mutagenesis to become lighter, darker or even different colors. We knew from our iGEM team that students really enjoyed making colored bacteria! For these reasons, we selected a chromoprotein expression project for our lab course.

2

Genes, Chromoproteins and Antisense RNAs

Use of this manual requires a little basic knowledge in chemistry and bio-chemistry. Such principles are well covered in many textbooks, including the first textbook on the principles of synthetic biology (Freemont and Kitney, 2012; 2nd edn. in 2015). The reader is encouraged to consult these textbooks for the structures of biological macromolecules and their functions such as base pairing and catalysis. Nevertheless, it is difficult to find practical information in three rapidly evolving fields central to our experimental design: codon bias, chromoproteins and antisense RNAs in *E. coli*. Before covering these three topics below, some basic aspects of molecular biology directly relevant to synthetic biology and our lab are introduced.

E. coli DNA: Chromosomes, Plasmids and Copy Number

E. coli is a bacterium growing in our colons and is the only "chassis" organism used in this manual.

Chassis

When the word "chassis" is used in synthetic biology, it simply refers to the organism that will be used to host the synthetic system. The most used and best characterized chassis of them all is *E. coli*. There are many different strains of *E. coli*, each having advantages and disadvantages. For cloning and assembly, cloning strains like DH5α are used because they give high transformation frequencies and good plasmid DNA preparations. However, for expressing synthetic devices, a healthier wild-type strain like MG1655 is more suitable than the rather weak and somewhat slower-growing cloning strains that give variable colony sizes. While many synthetic biology standards and assembly methods are chassis independent, others are not (e.g. promoters and codon bias). It is very important to bear this in mind when moving a gene from one chassis into another.

E. coli contains a single, double-stranded, circular, chromosomal DNA molecule of ~4.6 million base pairs encoding ~4400 genes. Some *E. coli* cells also maintain much smaller, double-stranded, circular DNA molecules of a few thousand base pairs known as plasmid DNAs (Fig. 2).

Plasmid DNA is easier to engineer than chromosomal DNA in many ways. For example, while the copy number of chromosomal DNA per cell is one, the copy number of a plasmid per cell may be anywhere from one to hundreds, depending on the particular origin of replication on the plasmid DNA. Thus, one way to increase expression of a gene on a plasmid is to increase the copy number of the plasmid, thereby increasing the number of gene copies per cell.

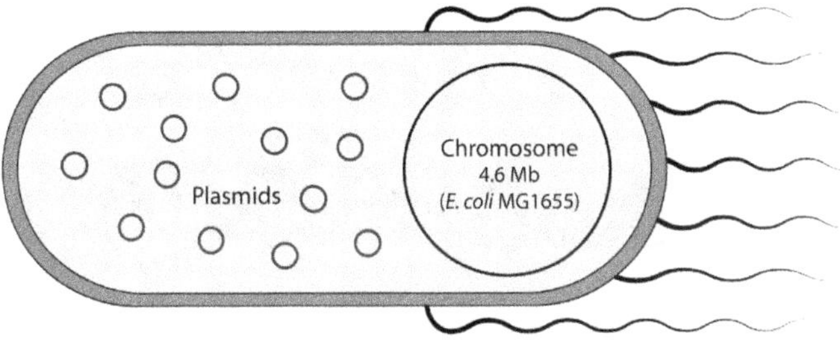

Fig. 2   *E. coli* cell containing chromosomal and plasmid DNAs.

Plasmid copy number and compatibility

A plasmid's host range, compatibility with other plasmids and copy number is determined by its origin of replication. Only one plasmid from each compatibility group can be stable in a strain at a time, so if you want to design a multiplasmid system (e.g. for our antisense system), the plasmids must come from different compatibility groups. The difference between expressing a certain construct from the bacterial chromosome (~1 copy), from a medium-low copy plasmid (~15 copies) or from a high copy plasmid (~500 copies) can be dramatic. To get high enough expression from a low or medium copy plasmid, a strong promoter may be required. However, this promoter on a high copy plasmid could give such high expression that it would be toxic to the cell.

Coupling of Transcription and Translation in Bacteria

The central dogma of molecular biology dictates that DNA encodes RNA which encodes protein (Fig. 3).

The particular type of RNA that encodes protein is called messenger RNA (mRNA). RNA synthesis from a DNA template is catalyzed by RNA

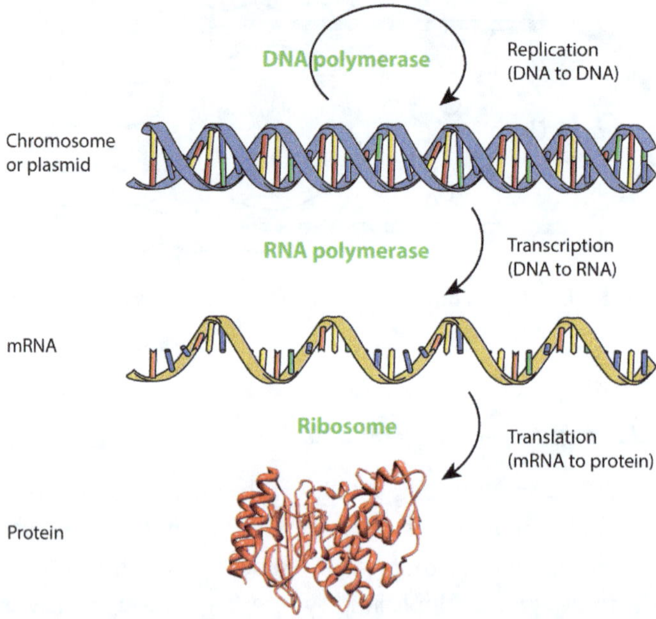

Fig. 3 The central dogma of molecular biology. The top and middle structures are taken from Wikipedia with permission.

polymerase and is termed transcription. Protein synthesis from an mRNA template is catalyzed by ribosomes and is called translation. In contrast to higher organisms, where transcription is physically separated from translation (due to splicing), bacteria couple transcription and translation (Fig. 4).

A practical implication for synthetic biology of this coupling is that translation can feed back on transcription very rapidly. So rapidly, in fact, that even the initial speed of the translating ribosome can determine whether or not the mRNA it is translating is synthesized completely or terminated prematurely (termed attenuation).

Some RNAs are not translated; these are termed stable RNAs or non-coding RNAs. Examples include the ribosomal RNAs (rRNAs) and transfer RNAs (tRNAs) involved in translation, and the small regulatory RNAs engineered as part of the lab course (see end of chapter).

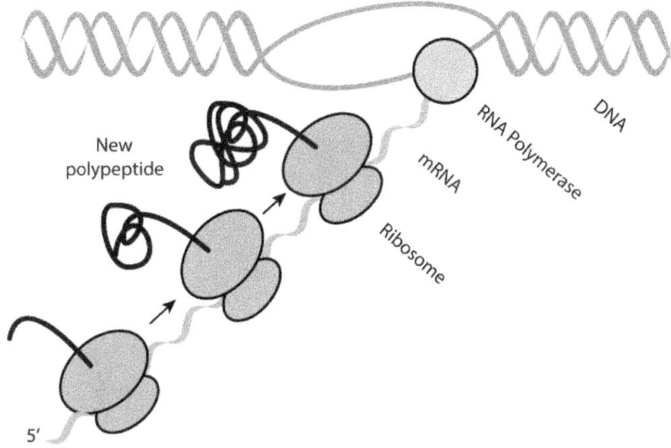

Fig. 4   Transcription and translation are coupled in bacteria.

Promoter and Terminator for Transcription

Initiation of transcription is determined by protein-DNA interactions. The proteins are called RNA polymerases and transcription factors. The DNA portion is called the promoter and lies at the upstream (5′) end of the gene(s) to be transcribed. In bacteria, one promoter may initiate transcription for one or several genes, an arrangement termed an operon (Fig. 5A).

In experiments described in this manual, expression levels of RNAs and proteins are often changed by using different DNA promoter parts, leaving the polymerase and transcription factors unchanged. In *E. coli*, the primary (housekeeping) transcription factor is called sigma-70. The complex between sigma-70 and RNA polymerase recognizes promoter sequences with the bipartite consensus shown (Fig. 5A). The TTGACA box lies at −35 nucleotides from the transcription start site (TSS) and the TATA (Pribnow) box lies at −10. The TATA sequence corresponds to one of the least stable double-stranded DNA secondary structures, thereby facilitating the opening (melting) of the double helix necessary for transcription initiation.

Another RNA polymerase often used in synthetic biology is from *E. coli* bacteriophage T7. The merits of T7 RNA pol include more efficient transcription, a single-subunit enzyme, and high specificity for a long

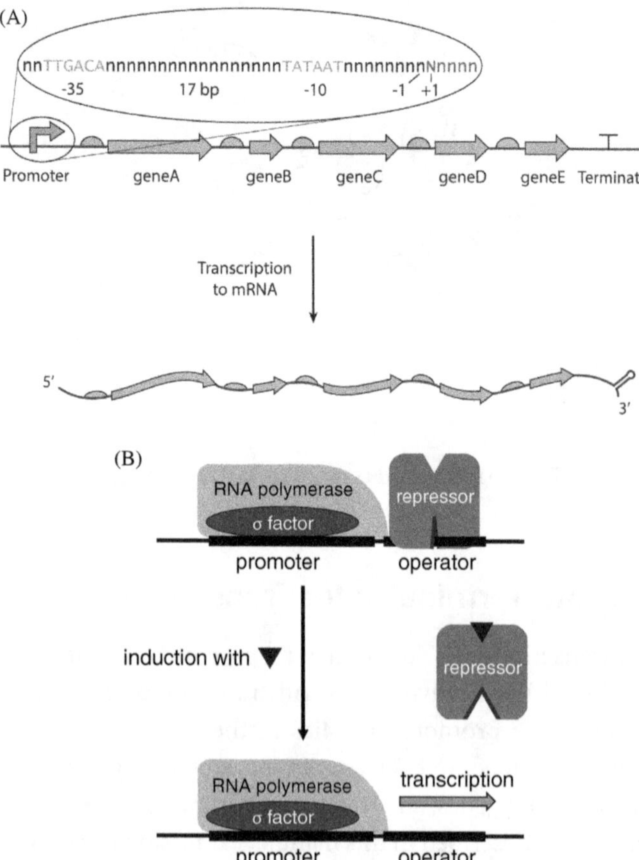

Fig. 5 (**A**) Consensus *E. coli* promoter sequence upstream of an operon encoding five genes. N is any nucleotide, and there is no nucleotide numbered 0. Semicircles represent ribosome binding sites (parts are represented according to the Synthetic Biology Open Language visual standard at http://www.sbolstandard. org/visual). (**B**) Regulation of transcription via an operator DNA sequence. Top: transcription of DNA by *E. coli* RNA polymerase is blocked physically by a repressor protein bound to the operator. Bottom: induction of transcription by a small molecule (triangle) that alters the conformation of the repressor to decrease its affinity for the operator.

(19-nucleotide) promoter sequence which allows construction of an independently working (orthogonal) transcription system in *E. coli*.

Additional protein-binding DNA sequences are often incorporated into the promoter region in synthetic biology to regulate transcription. Sequences called operators bind repressor proteins (Fig. 5B), while enhancer sequences bind activator proteins. The activities of both types of proteins can be controlled with small molecules called inducers. A commonly used combination is the lac operator/lac repressor/IPTG inducer. Indeed, the most efficient system for overexpressing proteins exploits the lac operator and T7 RNA pol (see Novagen's pET System Manual, 11th edn., beginning with their upper figure on p. 45; a His_6 tag is often appended to the amino or carboxyl terminus of the protein to enable easy purification on nickel beads).

Termination of transcription occurs when the polymerase synthesizes an RNA sequence containing a stem-loop structure immediately followed by a run of U bases (Fig. 6).

Termination structures facilitate dissociation of the nascent RNA transcript from its DNA template because of low stability of the DNA-RNA hybrid. Terminators are placed between genes to insulate them from each other by stopping the RNA polymerase from continuing to transcribe from

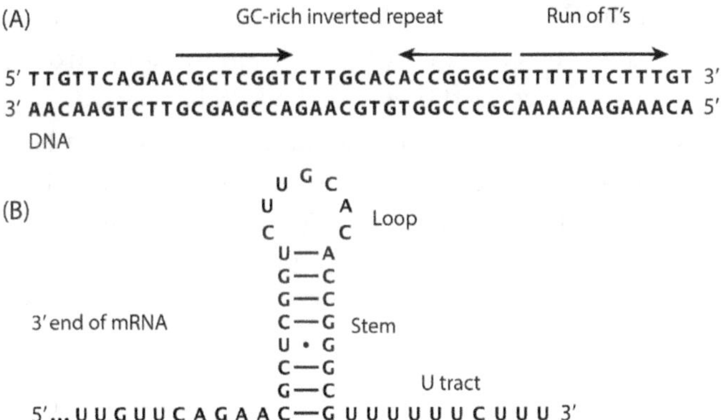

Fig. 6 **(A)** DNA sequence encoding a transcription termination signal. **(B)** Secondary structure of the terminator RNA sequence encoded in (A). Additional AU and GU base pairs are possible.

one gene into the next. However, some leakiness into adjacent genes is to be expected because terminators are not 100% efficient.

The presence of a terminator stem-loop structure also increases the half-life of the mRNA, which is another potential control point for synthetic biology. But in practice, altering mRNA stability is difficult to do in a predictable, modular manner. Nevertheless, it is important to bear in mind that mRNAs are generally degraded by ribonucleases very rapidly in bacteria, as this enables fine temporal control of gene expression by switching off transcription.

Ribosome Binding Site (RBS)

Initiation of translation in bacteria is primarily determined by base pairing. The first step is the binding of the small subunit of the ribosome to the RBS (also called the Shine–Dalgarno sequence), a sequence in the mRNA 5–8 bases upstream of the AUG start codon (Fig. 7A).

The RBS base pairs with the 3′ end of the 16S rRNA in the ribosomal small subunit (anti-Shine–Dalgarno sequence, ASD), with the efficiency of base pairing determining the efficiency of initiation of translation. This efficiency is dictated not only by the strength of the base-pairing hybrid but also by the availability of the RBS for base pairing. Single-stranded RBSs bind to the ASD most efficiently, but usually the RBS is sequestered somewhat by weak or strong base pairing in cis with nearby sequences in the mRNA (Fig. 7B). Thus, while there are several BioBrick™ RBS parts available, the efficiency of translation initiation from each will vary depending on the context of where the RBS is inserted.

The second step in translation initiation is base pairing between the start codon (almost always AUG) and the CAU anticodon of initiator fMet-tRNA^{fMet} (Fig. 7A). The efficiency of this step is not affected by mRNA sequences directly adjacent to the start codon, but it is decreased substantially by using an alternative start codon, GUG, which pairs less stably with the initiator tRNA anticodon.

Codon Bias

Once the fMet-tRNA^{fMet} is correctly positioned at the start codon of the mRNA by the ribosomal small subunit and accessory initiation factor

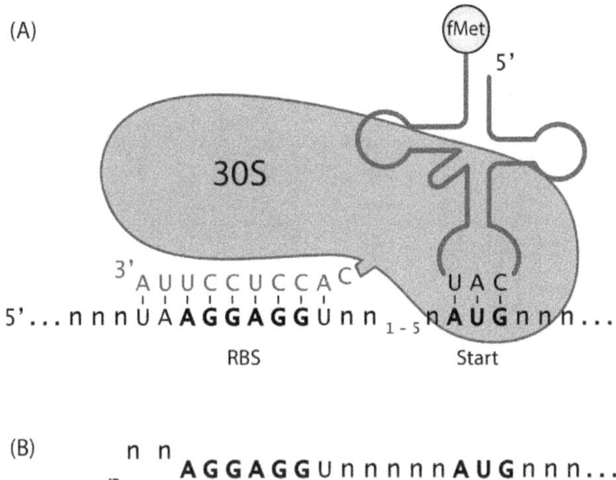

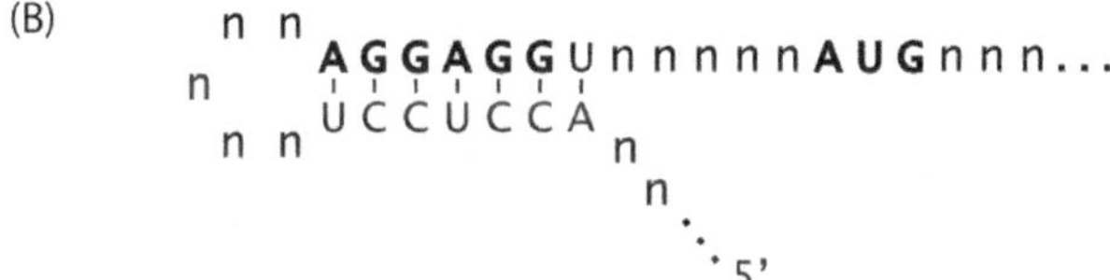

Fig. 7   Initiation of translation and inhibition of initiation. (**A**) For initiation of translation, first the RBS in the mRNA base pairs with the 30S subunit of the ribosome. Then a nearby downstream mRNA sequence termed the start codon base pairs with initiator tRNA charged with formylmethionine. For simplicity, initiation factors 1, 2 and 3 are not shown. (**B**) mRNA secondary structure involving the RBS inhibits binding of the 30S subunit and hence initiation.

proteins, the large (50S) ribosomal subunit binds to form the 70S ribosome. The ribosome then catalyzes translation of the next triplet codon into its cognate amino acid via base pairing with a cognate aminoacyl-tRNA. That amino acid is added onto fMet to form a nascent peptide, and this peptide is then extended one amino acid at a time until the ribosome reaches a stop codon and terminates translation. There are only 20 different, proteinogenic, elongator amino acids encoded by 61 different codons in the universal genetic code, so most of the 20 amino acids are encoded by more than one codon (termed synonymous codons; Fig. 8).

Interestingly, different codons for the same amino acid are used non-randomly in genes: there is always codon usage bias within synonymous codons, termed codon bias.

		2nd base				
		U	C	A	G	
	U	Phe	Ser	Tyr	Cys	U
		Phe	Ser	Tyr	Cys	C
		Leu	Ser	Stop	Stop	A
		Leu	Ser	Stop	Trp	G
5' base	C	Leu	Pro	His	Arg	U
		Leu	Pro	His	Arg	C
		Leu	Pro	Gln	Arg	A
		Leu	Pro	Gln	Arg	G
	A	Ile	Thr	Asn	Ser	U
		Ile	Thr	Asn	Ser	C
		Ile	Thr	Lys	Arg	A
		Met	Thr	Lys	Arg	G
	G	Val	Ala	Asp	Gly	U
		Val	Ala	Asp	Gly	C
		Val	Ala	Glu	Gly	A
		Val	Ala	Glu	Gly	G

Fig. 8   The universal genetic code.

Codon bias varies dramatically between different organisms and organelles, and even varies at different segments within one gene. Explanations for codon bias are varied and still being revised (Forster, 2012; Goodman *et al.*, 2013). However, there is no doubt that codon bias was selected by multiple different evolutionary forces such as mutational bias, GC richness of DNA, RNA secondary structure, and efficiency of translation. In fast-growing bacteria such as *E. coli*, translational speed is likely the main driver of codon bias. Genes from foreign organisms (e.g. chromoprotein genes) often express poorly in *E. coli* because some of the required cognate aminoacyl-tRNAs are relatively less abundant in *E. coli* than in the natural host. Indeed, expression can often be improved by adding tRNA genes to *E. coli* or by changing the codon bias of the coding sequence to that of *E. coli*. For the latter engineering approach, a table of codon bias in the most highly expressed proteins (Fig. 9) is more useful than a table of the overall codon bias of the *E. coli* chromosome (not shown).

		2nd base			
		U	**C**	**A**	**G**	
5' base	**U**	7.92 **UUU** F phenylalanine 23.25 **UUC** F 2.73 **UUA** L leucine 4.27 **UUG** L	16.33 **UCU** S serine 11.68 **UCC** S 1.98 **UCA** S 2.51 **UCG** S	6.72 **UAU** Y tyrosine 16.52 **UAC** Y 4.18 **UAA** stop 0.00 **UAG** stop	2.76 **UGU** C cysteine 3.81 **UGC** C 0.19 **UGA** stop 7.03 **UGG** W tryptophan	U C A G
	C	3.86 **CUU** L leucine 4.09 **CUC** L 0.82 **CUA** L 60.75 **CUG** L	4.38 **CCU** P proline 1.09 **CCC** P 5.18 **CCA** P 28.82 **CCG** P	6.78 **CAU** H histidine 14.21 **CAC** H 7.01 **CAA** Q glutamine 27.28 **CAG** Q	43.82 **CGU** R arginine 20.59 **CGC** R 0.67 **CGA** R 0.62 **CGG** R	U C A G
	A	15.79 **AUU** I isoleucine 43.86 **AUC** I 0.52 **AUA** I 21.67 **AUG** M methionine	20.64 **ACU** T threonine 26.70 **ACC** T 2.61 **ACA** T 4.17 **ACG** T	5.61 **AAU** N asparagine 29.21 **AAC** N 55.01 **AAA** K lysine 17.22 **AAG** K	2.19 **AGU** S serine 9.31 **AGC** S 0.63 **AGA** R arginine 0.03 **AGG** R	U C A G
	G	43.18 **GUU** V valine 7.67 **GUC** V 22.31 **GUA** V 14.98 **GUG** V	39.49 **GCU** A alanine 11.81 **GCC** A 24.87 **GCA** A 24.11 **GCG** A	19.27 **GAU** D aspartic acid 33.74 **GAC** D 57.86 **GAA** E glutamic acid 16.97 **GAG** E	45.55 **GGU** G glycine 34.17 **GGC** G 1.26 **GGA** G 2.36 **GGG** G	U C A G

3' base is the rightmost column (U C A G repeated for each block).

Fig. 9   Codon usage frequencies (1×10^{-3}) in abundant proteins in *E. coli*. The standard genetic code format is used and initiator codons are omitted. Adapted from Forster (2012).

Gene synthesis companies and free websites offer the use of programs for codon optimization (as well as DNA and mRNA structure optimization), but the rules are still unclear. In theory, protein expression levels should only be affected by the rate of initiation, not the rate of elongation. So, the only codons needing optimization should be the initial ones, as only they affect the rate of clearance of the ribosome from the initiation region of the mRNA. Yet codon optimization is usually performed across the entire gene. This latter approach is probably necessary for massive overexpression to prevent ribosome "traffic jams" throughout the mRNA, but is it important for most synthetic biology applications which do not involve massive overexpression?

Chromoproteins

In nature, many organisms have bright colors caused by a wide spectrum of colorful pigments. While many of these pigments are small molecules produced by metabolic pathways, some of these colors come from pigmented proteins, chromoproteins. Most of these proteins get their color via a pigmented prosthetic group, e.g. heme bound in cytochromes or hemoglobin. Here, we focus on a different type of proteins, the family of fluorescent proteins and chromoproteins that are common in corals, sea anemones and jellyfish (Dedecker *et al.*, 2013). The most famous protein from this class is **green fluorescent protein (GFP)** from the jellyfish *Aequorea victoria*. These proteins do not require any prosthetic groups, but instead, the **chromophores** are formed through reactions between side chains of the protein itself (S65, Y66 and G67 in the case of GFP; single-letter abbreviations of the amino acids are given in Fig. 9) in a form of post-translational oxidation called maturation. As GFP-type proteins can fold and mature properly in almost any organism without exogenous cofactors and they are encoded by single, small genes (~700 bp), they have become very useful as reporters in molecular and synthetic biology. The fluorescent proteins have the advantage of high sensitivity, while the chromoproteins enable instrument-free detection by eye.

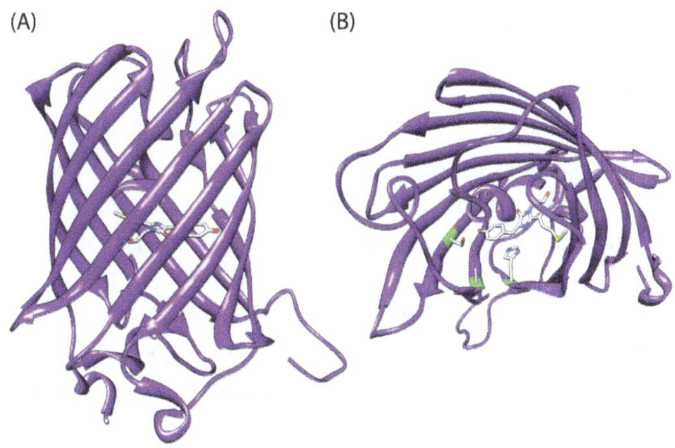

Fig. 10   Three-dimensional structure of chromoprotein asFP595 (also called asCP, KFP or asPink). (**A**) View illustrating beta-barrel structure with the covalently-bound chromophore rigidified in the center. (**B**) View from a different angle that better shows the chromophore. Amino acid residues that affect the color of the chromophore are marked with green (A148, S165 and H203; numbering based on GFP). Rendering based on PDB 1XMZ (Quillin *et al.*, 2005).

GFP-type proteins all share a similar beta-barrel structure, with the covalently bound chromophore protected inside the barrel (Fig. 10).

The color of the protein is dependent on which amino acid residues form the chromophore as well as the interactions between the chromophore and the neighboring amino acid side chains inside the barrel. Many natural GFP-type fluorescent proteins have been engineered through both directed and random mutagenesis to generate a large number of novel proteins with new spectral properties, improved fluorescence, higher photostability, lower toxicity, increased pH tolerance and improved folding and faster maturation (database at https://www.fpbase.org/). The fluorescent proteins can have a wide range of colors: cyan, blue, green, yellow, orange and red. GFP-type chromoproteins mature slowly, require overexpression for visualization, are somewhat toxic to *E. coli*, and most are yet to be engineered (Alieva *et al.*, 2008; Liljeruhm *et al.*, 2018). Ultimately, we overcame toxicity problems through engineering mRFP1 (Bao *et al.*, 2020).

Small Regulatory RNAs (sRNAs)

Regulation of gene expression in bacteria is often done post-transcription-ally via sRNAs, antisense RNA molecules that either increase or decrease mRNA translation. A typical sRNA is 50–400 bases long, and one sRNA can regulate several different mRNA targets. Many sRNAs are dependent on an RNA binding protein called Hfq that can both protect the sRNA from degradation and facilitate the interaction between the sRNA and its mRNA target (Sharma *et al.*, 2011).

Repression of gene expression via sRNAs usually involves binding to the RBS or the start of the coding region on the mRNA, hindering the ribosome from binding and thereby blocking initiation of translation (Fig. 11A).

There are also cases where the sRNA binds elsewhere on the target mRNA, and instead promotes degradation by recruiting ribonucleases (Fig. 11B). In cases where an sRNA activates translation of a gene, the sRNA often binds to the 5′ untranslated region of the mRNA to open up a structure that hid the RBS from the ribosome (Fig. 11C).

In synthetic biology, sRNAs show great potential for rapid fine-tuning of gene expression (Na *et al.*, 2013) compared with standard approaches. For example, if there is only a single gene copy, a genetic knockout might be lethal and would not allow partial inhibition of the gene. And com-pared to transcriptional regulation via repressors, translational regulation via sRNAs has much shorter response times. A small RNA can start repressing its target immediately after transcription, while a repressor gene needs to be translated to a protein after transcription. Even then, the expression of the gene will continue as long as there is still mRNA remain-ing in the cell. sRNAs are also interesting tools for gene regulation since they are generally smaller and easier to construct than protein repressors. A common approach when designing synthetic sRNAs is to use the Hfq-binding portion of natural sRNAs and then engineer the 5′ antisense por-tion to bind the target mRNA (see Fig. 35 and top left of Fig. 50). This can be done either through rational design and secondary structure modeling or by random mutagenesis and screening. When designing synthetic sRNAs, it is important to keep their target specificity in mind. For exam-ple, RBSs tend to be quite similar to each other, so a synthetic sRNA com-plementary to an RBS would bind non-specifically to many different mRNAs, likely being toxic to the cells.

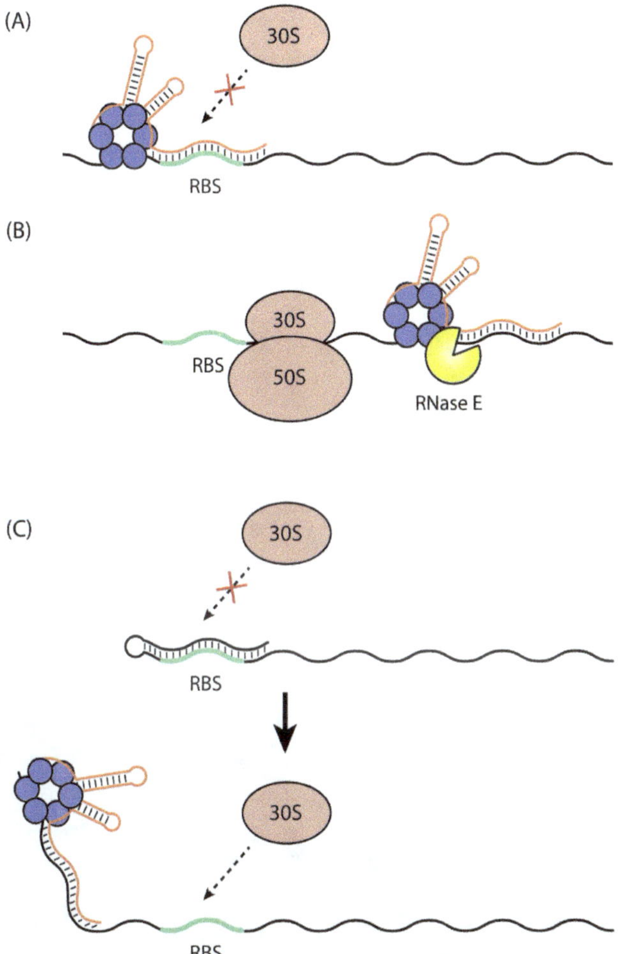

Fig. 11   Mechanisms of action of sRNAs. **(A)** An sRNA blocking initiation of translation by base pairing with the RBS on the mRNA. The blue circles represent Hfq protein. **(B)** An sRNA binding downstream on the mRNA, promoting degradation of the mRNA by recruiting ribonuclease. **(C)** Top: The 5′ untranslated region of the mRNA can contain structures that sequester the RBS from the ribosome. Bottom: Binding of an sRNA to the 5′ untranslated region of the mRNA opens up the RBS so that the ribosome can bind and initiate translation.

3

Lab Rooms and Equipment

The Physical Lab Spaces

Our theory and lab course in synthetic biology in 2013 catered for 27 undergraduate students at Uppsala University. Lab work was mainly conducted in a single room measuring 8 × 13 meters, containing 40 meters long of discontinuous bench space, easy-to-clean surfaces and chairs, and standard safety features (Fig. 12). Identical rooms were used by our iGEM teams of up to 26 students per summer.

During the late stage of the COVID-19 pandemic, the course was permitted with the following precautions: screens were erected between each side of the benches, the student density was halved, everyone wore masks, and people with symptoms or recent COVID-19 contacts stayed home.

Fig. 12   Main lab room for 27 students.

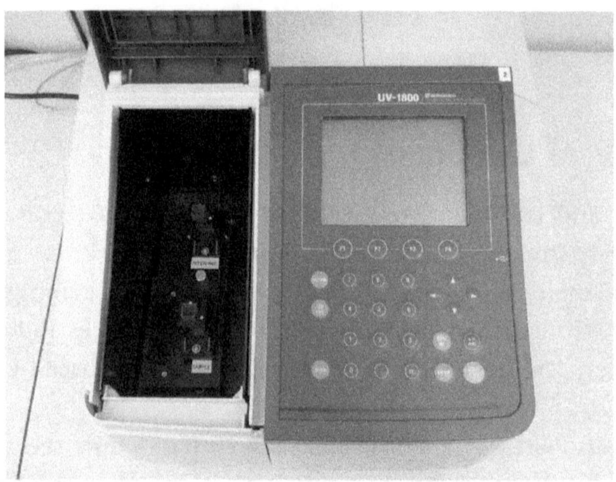

Fig. 13   UV-visible spectrophotometer containing 1 mL sample and reference cuvettes. Instruments typically measure absorbances of light wavelengths between 220 and 750 nm. Disposable plastic cuvettes are suitable only for the visual region (>380 nm) due to absorbance of UV by plastic. Quartz cuvettes are required for the UV region.

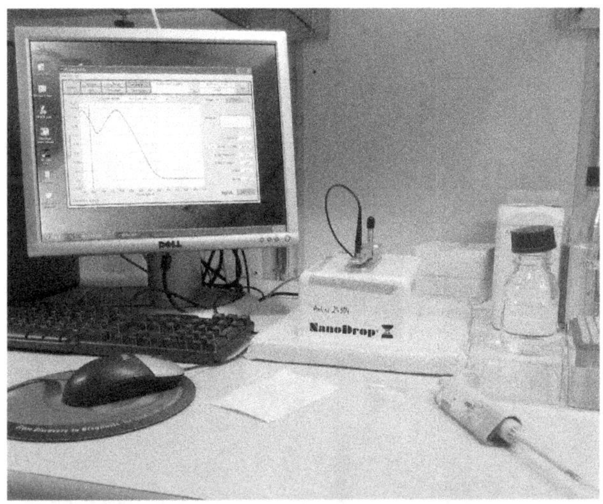

Fig. 14 Nanodrop™ (see also BioDrop) UV-visible spectrophotometer for μL samples. Compared with spectrophotometers for 1 mL cuvettes, we found Nanodrop™ to be more sensitive (6 ng/μL). Furthermore, it has the advantages of not requiring dilution of each sample before measurement and not requiring special cuvettes.

Adjacent to the main lab room was a room containing spectrophotometers (Figs. 13 and 14) and large centrifuges, a room for storage and weighing of chemicals, and an autoclave machine room.

Because of safety regulations forbidding eating and drinking in the labs, we had adjacent rooms for coffee/tea, lunch and group study.

Though we had two teaching assistants and two main lab rooms assigned for the class, we found it more practical to have all 27 students filling one lab room to capacity instead of spreading between two rooms. This halved the number of introductions required, ensured that all students received the same information simultaneously, and facilitated maintaining at least one teaching assistant in the lab room as a safety measure. The class was divided into eight groups of three or four students, and the students and equipment were spaced to minimize congestion. Thus, PCR machines (Fig. 15), gel equipment (Figs. 16 and 17), etc., were each spread into multiple stations.

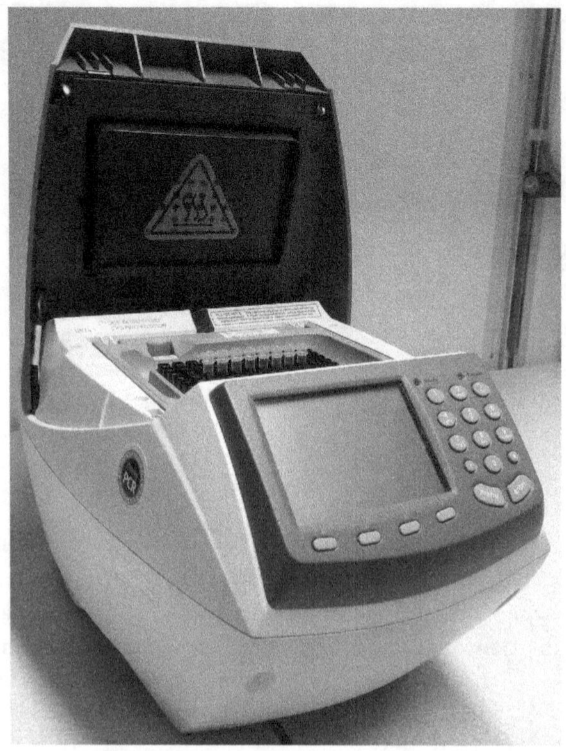

Fig. 15   PCR thermocycler. Machines contain metal blocks that can be programmed to undergo rapid cycles in temperatures between 4 and 99°C.

Equipment

This lab course was designed to take advantage of standard teaching lab equipment to keep the budget low. **All required major and minor items are listed in the Appendices** and special safety considerations are given in Chapter 4. Some instructions for students are given here:

Autoclave Machine

Because some bacteria survive boiling, sterilization of solutions requires heating aqueous solutions for 20 min at temperatures well above 100°C (typically 121°C). In order to prevent the aqueous solutions evaporating at

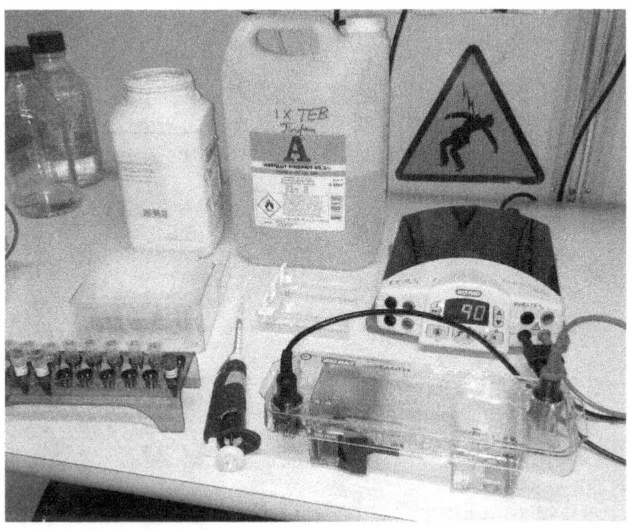

Fig. 16   Agarose gel electrophoresis. A high voltage is applied across a horizontal gel submerged in buffer so that nucleic acids migrate toward the positive electrode at speeds roughly proportional to their sizes.

Fig. 17   UV light box enables imaging of DNA bands in an agarose gel stained with Sybr®Safe (or ethidium bromide).

Fig. 18 Autoclave containing bottles with loosened autoclavable plastic caps. The machine generates steam under high pressure to kill bacterial contaminants.

such high temperatures, a steam atmosphere at high pressure is required, so water is initially added to the autoclave chamber (Fig. 18).

- Do not handle the autoclave machine without being instructed to do so.
- Only use autoclavable plastic, as other plastic can melt.
- Use distilled or double-distilled water to avoid lime deposits.
- Fill up only to 80% of the volume of a bottle, except for agar media for plates, which are filled up to 60% of the bottle.
- Loosen the cap of the bottle.
- Liquids are autoclaved for 20 min.
- Use insulated gloves when unloading.

Burners

These are essential for microbiology, both for sterilization and for preventing bacterial contamination. Burners cause upward air motion, thereby decreasing contamination of media. Great caution is required because the flames can be almost invisible and because of the danger of burns and fires.

- Use the correct gas container for the burner.
- Do not use plastic gloves while working near a flame.

- Be careful with glass spreaders and wire inoculation loops as both are fragile.
- See Chapter 4 (Fires section and Fig. 20) for additional safety advice on burners.

Gel Equipment

- Use gloves at all times because the equipment has been in contact with ethidium bromide and/or Sybr®Safe dyes.
- Be cautious when handling melted agarose — it is very hot.
- Add the comb straight after pouring the melted agarose into the tray.
- Make sure that the gel solidifies on a level surface.
- To prevent electric shock, do not apply voltage until the lid is properly placed upon the buffer chambers (Fig. 16).

Centrifuges

- If the centrifuge has a cooling system, the lid must not be left open or water will condense inside — treat it like your refrigerator!
- Use adaptors or a special rotor if centrifuging tubes for smaller volumes than 1.5 mL.
- Always balance the centrifuge rotor!
- Never leave the centrifuge until it has reached maximum speed — it should be shut down immediately if there is a loud noise!

Micropipettes

The micropipettes (or pipettemen; Figs. 14, 16) are key tools and it is important to keep them calibrated and use them properly.

- Never turn the knob past the printed volume range for that pipette. Consult the manufacturer's instructions for the exact range of your pipette.
- The largest error rate lies at the bottom of a pipette's range.
- Use the correct pipette tips for the volume being pipetted.
- Hold the pipette vertically while pipetting to ensure that roughly the right volume is aspirated (you will learn what looks right from experience).

- Aspirate the liquid carefully for accuracy and to avoid contamination of the pipette. This especially applies to cultures and non-viscous solvents like ethanol.
- Make sure that all of the volume leaves the tip.
- If a micropipette becomes contaminated or its knob is turned too far, give it to the lab assistant for cleaning or calibration.

Laboratory Bench

- Ensure that the table does not get too overcrowded with papers.
- Do not have any paper on the shelf over the lab bench as it is a potential fire hazard when using the burner. Never use paper bench coat when using a burner!
- Wash the bench with 70% ethanol prior to, and after, any work with cells and media.
- Clean the bench at the start and end of the day to avoid accumulation of paper and equipment that could be a source of bacterial contamination.

4

Safety is Priority #1

The top priority in any lab must be safety. The risks are not insignificant, as most seasoned scientists have either directly witnessed lab accidents or heard of their occurrence in their own institution. Risks can be minimized by understanding hazards and safety procedures, so safety training by certified personnel is mandatory before commencing lab work. Know where safety equipment and first aid kits are located. If an accident occurs, notify the lab assistant immediately.

Fires

Everyone should become familiar with institutional fire regulations.

Know the locations of the fire-fighting equipment, fire alarms and evacuation routes closest to the lab. A small fire can be extinguished quickly by smothering it in a fire blanket or by spraying it with a fire extinguisher. There is usually a choice of two kinds of fire extinguishers and the

Responding to Fires

The usual order of action is:

1. Save lives.
2. Call the fire brigade.
3. Alert people in the area.
4. Extinguish the fire if possible.
5. Close doors to the area.
6. Evacuate.
7. Re-assemble outside the building at the designated meeting point.

simplest choice is the one containing carbon dioxide, i.e. the one with a large nozzle (left of Fig. 19).

The carbon dioxide extinguisher is recommended for all types of fires, whereas the alternative liquid/foam fire extinguisher should not be used on electrical fires because the liquid stream can conduct electricity back to the firefighter!

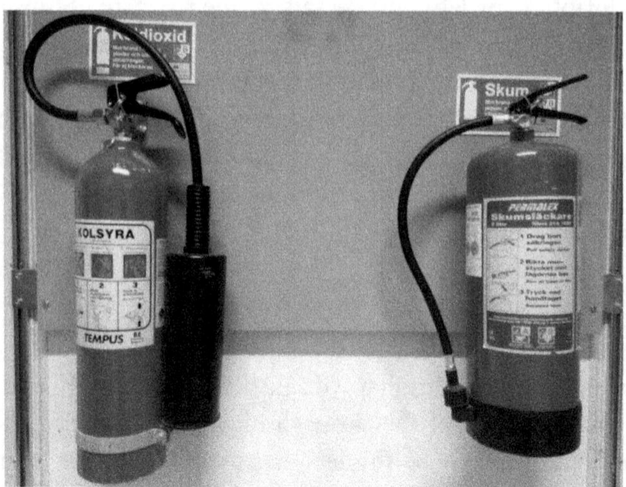

Fig. 19   Fire extinguishers containing carbon dioxide (left) and liquid/foam (right).

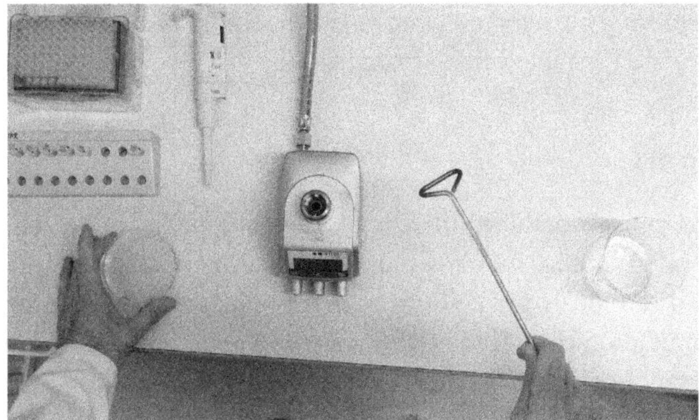

Fig. 20   Safe set-up for plating cell cultures using a burner. Precautions include *not* wearing plastic gloves.

Lab fires in biological labs are caused most commonly by the plating of cell cultures (Fig. 20).

Typically a metal or glass spreader is dipped into a glass container of ethanol. The ethanol on the spreader is then ignited by passage through a burner flame, and burning drops of ethanol inadvertently fall from the spreader either (i) onto paper bench coat covering the bench or (ii) into the glass container of ethanol, with the container possibly cracking due to the heat of the fire.

To prevent this, one should:

1. Never use paper bench coat near burners.
2. Use a low reservoir volume of ethanol.
3. Place the ethanol reservoir at least a foot from the burner and on the opposite side of the burner from the plates (Fig. 20).

Other burner safety tips are:

4. Never wear plastic gloves when working with a flame.
5. Carry the gas container for the burner carefully with a firm grip. If it is dropped, make sure that there is no leakage by smelling for gas.
6. Notify the instructor of any gas leakage and do not light burners when there is a smell of gas.
7. Always light the match before opening the gas valve.

8. Avoid placing burners too close to overhanging shelves.
9. Never leave the table while a burner is on.

Chemicals

On each chemical container, there is a label that specifies the potential danger of the substance for humans and/or the environment. Chemicals should be handled cautiously with gloves, both for your safety and for decreasing the contamination risk. Always wear a lab coat and shoes as additional protection. Handling of liquid nitrogen requires a face mask and cryogenic gloves in a well-ventilated area, not a closed small room (Fig. 21).

For handling eye hazards such as hydrochloric acid, hydroxide solutions and organic solvents, safety glasses are required, and, depending on the quantities, also a fume hood (Fig. 21). This lab course is designed to minimize the use of such hazards and fume hoods (e.g. phenol is not required). Radioisotopes are also not used.

Upon chemical spillage, check your clothes and lab coat. Read the signs on the chemical container and the Material Safety Data Sheet (MSDS; available online) for further direction. Wipe off and wash the skin and notify the lab assistant. Know where the emergency eye wash and shower stations are and flush eyes immediately after contamination. In laboratories, it is not permitted to mouth pipette, drink, eat or even chew gum. Storage and disposal of food and drink is also forbidden. Wash your hands when leaving the lab to prevent contamination of your face and food.

Ethidium Bromide

This dye is a potent mutagen that is widely used as a sensitive DNA stain in agarose gel electrophoresis. Its properties derive from its planar structure (Fig. 22) having a propensity to intercalate into double-stranded nucleic acids.

Because it is probably the most toxic chemical used in this manual and it requires special disposal procedures at most institutions, we strongly recommend the safer substitute, Sybr®Safe, for all labs (Fig. 17). The relative sensitivities of these two DNA stains are very similar. Always wear

Fig. 21   Safe pouring of liquid nitrogen. If extra stability is needed, the bucket can be wedged in a corner. In the background is a fume hood for acids and solvents.

nitrile gloves when handling solutions of these stains, as vinyl gloves are more permeable.

Biological Safety and Disposal

Whatever you wish to call the preparation of new organisms by molecular methods (GMOs), e.g. synthetic biology, recombinant DNA, molecular

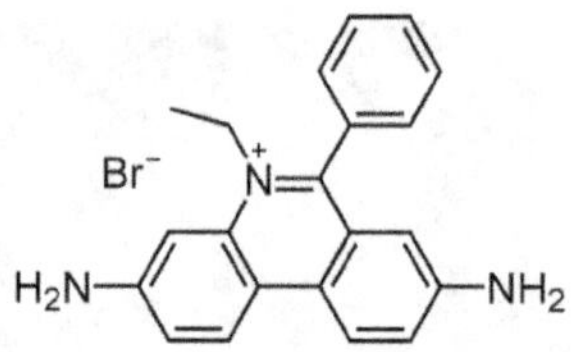

Fig. 22   Chemical structure of ethidium bromide dye. GelRed is marketed as a less-toxic, dimeric derivative of ethidium bromide. Taken from Wikipedia with permission.

cloning or genetic engineering, the field is regulated in most countries according to international biosafety guidelines:

1. CDC, Atlanta. Biosafety in Microbiological and Biomedical Laboratories. http://www.cdc.gov/biosafety/publications/bmbl5/bmbl5_sect_iv.pdf
2. WHO, Geneva. Laboratory Biosafety Manual, 2004. http://www.who.int/csr/resources/publications/biosafety/Biosafety7.pdf
3. ECDC, Directive 2000/54/ec of the European parliament and of the Council of 18 September 2000 on the protection of workers from risks related to exposure to biological agents at work (seventh individual directive within the meaning of Article 16(1) of Directive 89/391/EEC. http://eur-lex.europa.eu/LexUriServ/LexUriServ.do?uri=OJ:L:2000:262:0021:0045:EN:PDF

When working with microorganisms such as bacteria and viruses, there are four BioSafety Levels (BSL) numbered BSL1–4. This lab course is designed to be at BSL1, the lowest risk level. (The next lowest level, BSL2, involves pathogens that can, with low probability, cause moderate harm to people working with them, thus requiring special training.)

The lab course includes expression of chromoproteins and antibiotic resistance proteins from plasmids transformed into well-categorized cloning strains of *E. coli*, e.g. DH5α. These cloning strains have low fitness outside the laboratory and are non-pathogenic.

BSL1 requires decontamination of disposables, such as used tips and tubes, through steam autoclaving before disposal. Contaminated LB or SOB media must not be poured directly into the sink as they are environmental hazards. After cell growth, cultures and culture waste products are sterilized with 0.01% iodine or dilute bleach for a couple of hours before sending the flasks for cleaning. Glass pipettes are sterilized by submersion in Virkon disinfectant solution for a couple of hours before rinsing and washing. Disposables such as tubes, pipette tips and gloves are sterilized by autoclaving before disposal. Sharp objects, such as needles and broken glass, are handled carefully and disposed of in a puncture-resistant plastic container. There are strict rules to always wear laboratory coats when located inside the lab area. Protective eyewear is used when necessary, e.g. when there is risk for splashes or from exposure to artificial UV radiation. Protective gloves must be changed upon contamination or if otherwise necessary. Before leaving the lab area, the workbench should be sterilized, and hands washed.

Dangerous Equipment

Electric shocks can occur from mishandling equipment for agarose or polyacrylamide gel electrophoresis due to the high voltage differential between two exposed buffer reservoirs. Most modern commercial gel equipment has a built-in safety feature whereby attachment of the electrodes requires concomitant covering of the buffer reservoirs (Fig. 16). However, older or "home-made" equipment may require labeling as hazardous. The simultaneous touching of both buffer reservoirs with different hands during electrophoresis must be avoided!

Another danger with gels can occur when heating and cooling agarose solutions/gels. Always use conical flasks, not sealed bottles, because the latter may explode! And electrophoresis at too high a voltage will melt the gel.

Centrifugation creates enormous forces on rotors, especially at the highest speeds and when rotors are imbalanced. Rotors also eventually develop metal fatigue and can crack. For these reasons, never leave a centrifuge until it has reached maximum speed so that you can turn it off as soon as it begins to make a strange noise. This is important not only for safety, but also for preserving the equipment. As a general rule, dangerous and/or expensive equipment such as centrifuges, electrophoresis power packs, autoclaves and spectrophotometers should not be used without proper instruction.

5

Lab Course Projects

Time and Resources

As outlined in the introductory chapter, the goal of the lab course is to teach key synthetic biology technology and also to give students the opportunity to create their own projects with varying difficulties. Our ongoing course detailed in this manual is up to five full-time weeks of laboratory work (Table 7 in the Appendices). Each day begins with a one-hour lecture or tutorial on synthetic biology theory, followed by the rest of the day in the lab.

While our schedule works well according to the anonymous student evaluations, there is considerable flexibility in the protocols that allows adaptation to alternative scheduling requirements. Synthetic biology foundational technology has simplified molecular biology so famously that novices, even high school students and do-it-yourselfers/biohackers, can now expect to succeed in molecular cloning on first attempt within a week.

As few as one-and-a-half more weeks are sufficient for designing and building altered gene functions with a reasonable chance of success. Longer times provide for trouble-shooting and also encourage more creativity and self-directed learning. Should more than five weeks or alternative projects be desired, other experimental suggestions are given at the end of this chapter. All protocols can be completed, or at least taken to a potential storage point, within half a day, adding further flexibility. Although bacterial cultures and some PCR reactions require overnight incubations, these incubations are performed unsupervised, and their main requirement is starting them at the end of the day.

Space and major equipment requirements for 27 students were given in Chapter 3. The quantities of equipment, disposables, chemicals and molecular biologicals are given in Tables 2–6 in the Appendices and are calculated to fit 27 students working for five weeks.

Project Overview and Learning Objectives

A central feature of synthetic biology is engineering the function of genes in a manner that is seen at the organismal level. One of the most dramatic and recent illustrations of this is to color bacteria by expressing a chromoprotein, the first step of this course. The aim of the remainder of the course is to up- and down-regulate the protein expression level in a rationally designed way. The main methods are BioBrick™ DNA cloning and PCR mutagenesis. At the end of the lab course, students should be able to:

- understand basic principles in synthetic biology;
- perform experimental protocols central to synthetic biology;
- use the scientific method to generate hypotheses, design experiments to test them, and analyze the results;
- document and present clearly their methods, results and conclusions.

The Lab Notebook

Laboratory notebooks always have been, and still are (despite the advent of personal computers), essential for research. For this reason, we recommend that student lab books be assessed for a portion of student grades (see The "Dreaded" Exam at the end of this chapter). At a minimum, the lab book documents methods in sufficient detail that they can be repeated by you and others (a cornerstone of science), documents results so they can be analyzed (and hopefully published!), and documents stored samples so they can be used with confidence. Documentation should be clear enough that scientists in your field at your institution can understand it without you explaining it.

A proper lab notebook has numbered pages that cannot be removed. If there are no page numbers on your book, number each page now. All entries should be written in ink, not pencil, so they cannot be erased. On the outside cover, write your name, book number and laboratory. Inside the front cover, provide contact information in case the book is lost or stolen. Write "Contents" at the top of the first two pages and then proceed to write up your experiments sequentially from front to back. Typed protocols can be pasted at the back. An important tip is to always write directly into your notebook and not to keep a second notepad on the side for drafts. By following this tip, potentially important information will not be lost and you will not lose time (and perhaps accuracy!) in copying it into your lab book. Any mistakes should be crossed out but remain readable. It is important that you do not leave any big open patches — cross out such patches before moving to the next page. Adding data in retrospect or erasing, removing or destroying data is prohibited. This can be important for fraud, patent or even Nobel Committee investigations!

In your lab notebook, write down everything you plan, do, notice and analyze throughout the day in the lab. Everyone writes their own lab notebooks in their own way, but there must be a structure to it. For every experiment, include headings in the following order:

- Title (at the top of the page)
- Date (in unambiguous international YYYY-MM-DD format) in the left-hand margin
- Aim or Hypothesis
- Methods
- Results
- Conclusions

Data, readouts, gel images, photos, etc., are to be properly attached in the lab notebook. Wherever possible, use original documentation (e.g. a machine readout with date and data rather than hand copying out the data). Anything that is impossible to store in a lab book (e.g. due to size) should be dated and referred to clearly in the lab book. Writing thorough Results and Conclusions will help you support or refute your hypothesis and develop new hypotheses. Forcing yourself to write down detailed conclusions often leads to ones that were not apparent when looking at all the data at once. Ideally start with discussing negative and positive controls: if either failed, your experiment is probably a wash! Then go over all other gel lanes, agar plates, etc., one by one.

Lab Section 1. Preparation of Chemical Solutions and Agar Plates

Protocol 1 (Chapter 6). In order to perform the experiments, all six chemical solutions and the plates listed in Protocol 1 are required. Schedule preparation of the six solutions on the first day and the agar plates on the second day because of the autoclaving delay (Table 7 in Appendices). To avoid several groups competing for the same chemical stocks, the lab teacher can specify different orders of preparation of solutions for different groups. Alternatively, if time is severely limited, teachers could prepare the solutions for the students beforehand. The same applies to LB antibiotic agar plates; teachers may supply LB plates for initial or all rounds of cloning. Some labs have a media facility that provides agar plates. As soon as plates are in hand, practice **Protocol 7: Bacterial re-streaking techniques.**

The importance of preparing solutions correctly cannot be overemphasized. Trouble-shooting a failed experiment is hard enough without having to second guess whether or not the problem was due to an incorrectly made or contaminated solution. For this reason, it is crucial to document in your lab book the exact calculations used to make up your solutions. Otherwise, how can you check if there was a mistake? Use common sense regarding the accuracy of weights and volumes: for biological experiments, four significant figures are not required!

Lab Section 2. Coloring Bacteria by Adding a Promoter to a Chromoprotein Gene

Remember that **for protein expression, four DNA parts must be combined** (Fig. 5A):

- A promoter, for initiation of transcription.
- A ribosome binding site (RBS), for initiation of translation.
- A protein coding sequence (CDS or ORF), for translation elongation and termination.
- A plasmid vector, for stable maintenance and replication of the above three parts. (An alternative discussed in Chapter 7 is a chromosomal integration site.)

Fortunately for us, these four types of DNA parts have been standardized by synthetic biologists as BioBricks™, so they can be assembled easily (like Lego® blocks). The assembly process is plasmid DNA cloning using restriction endonucleases and DNA ligase, invented 50 years ago by Boyer and Cohen, but in a much simpler, standardized form (Knight, 2003) called **BioBrick™ assembly** (Fig. 23).

The most useful DNA parts are available from the Registry of Standard Biological Parts.

The Registry of Standard Biological Parts

The iGEM Foundation handles the parts registry, providing information about thousands of parts inserted in plasmids, almost all of which are BioBrick™ compatible. Annual iGEM distribution kits contain ~700 plasmids, including those for most workhorse enzymes in Appendix Table 6 and some parts in Table 1 (https://technology.igem.org/distribution/handbook#h-whats-included-in-the-2024-distribution). Their other BioBricks™ are no longer available, but we have made ideal lab course plasmids available from the non-profit company Addgene (Table 1).

BioBricks™ have a helpful nomenclature: http://parts.igem.org/Help:Plasmid_backbones/Nomenclature.

Each part is defined by a code beginning with **BBa_**. The plasmid backbone flanking each part is coded as **pSB#X#** (Shetty *et al.*, 2008), where:

- pSB = plasmid Synthetic Biology (i.e. a BioBrick™ vector)
- first number = origin of replication (e.g. 1 is high copy, 3 is low-medium copy)
- fourth letter = antibiotic resistance (e.g. C = chloramphenicol; this version is available for almost all inserts)
- second number = the generation of the vector

Most commonly, parts are inserted in pSB1C3 (Shetty *et al.*, 2011), a high copy plasmid vector. Sometimes vectors are called parts, and sometimes not. The nomenclature does not refer to the particular BioBrick™ assembly standard used, but it is usually RFC10 and the web page for each part in the registry shows which standards the plasmid is compatible with. For information on the five assembly standards supported by the registry, see http://parts.igem.org/ Help:Standards/Assembly.

For their first challenge, students will engineer protein expression by assembling the parts discussed above (promoter and fused RBS-CDS) into one destination plasmid.

Thus, the chromoprotein-expressing plasmid (Fig. 24A) only needs to be assembled from three different DNA restriction fragments (Fig. 24B):

1. promoter (upstream part)
2. RBS-chromoprotein CDS (downstream part)
3. plasmid vector (backbone part)

Successful in vitro assembly of the plasmid followed by transfer of the plasmid into bacteria (transformation) will start steady (constitutive) expression of the chromoprotein in vivo. This process will be directly visible as colorful bacterial colonies on the next day. We recommend that students are given different unknown chromoprotein genes as a starting point so they may practice the scientific method in determining which genes they have.

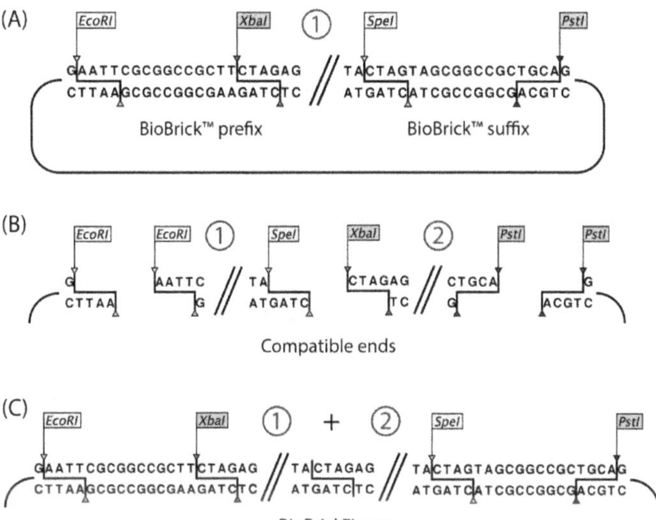

Fig. 23 The original and most commonly used BioBrick™ assembly standard, RFC10. (**A**) BioBrick™ plasmid containing part 1. The RFC10 standard uses the prefix shown or a shorter prefix (see below), the suffix and only four restriction enzymes, EcoRI, XbaI, SpeI and PstI, that cleave under the same conditions. Their restriction sites must occur only once per plasmid at the flanking locations shown. For coding region parts starting with ATG, the 3'-terminal AG of the prefix sequence is omitted to give more flexibility for the adjacent Shine–Dalgarno sequence (Fig. 7A; http://parts.igem.org/Help:Prefix-Suffix). These sequences also decrease the likelihood of methylation that would affect restriction digestion. (**B**) Digestions forcing cloning of part 1 upstream of part 2 into a destination plasmid vector. Alternatively, part 1 could be cut as shown for part 2, and part 2 could be cut as shown for part 1, to force clone part 1 downstream of part 2 (not shown). (**C**) Ligation of the fragments in **B** utilizes the compatible sticky ends (overhangs) of XbaI and SpeI to create a "scar" which cannot be cut with either enzyme. The end result is a new BioBrick™ plasmid containing a larger part, labeled 1 + 2, flanked by the standard prefix and suffix. For simplicity, NotI restriction enzyme sites are not shown.

The plasmid construction begins, not with the three restriction fragments themselves, but with three cell strains, each containing one of the three different DNA plasmids (top of Fig. 25):

1. Plasmid containing promoter insert BBa_J23110 bounded by BioBrick™ restriction sites (upstream part). It should be mentioned

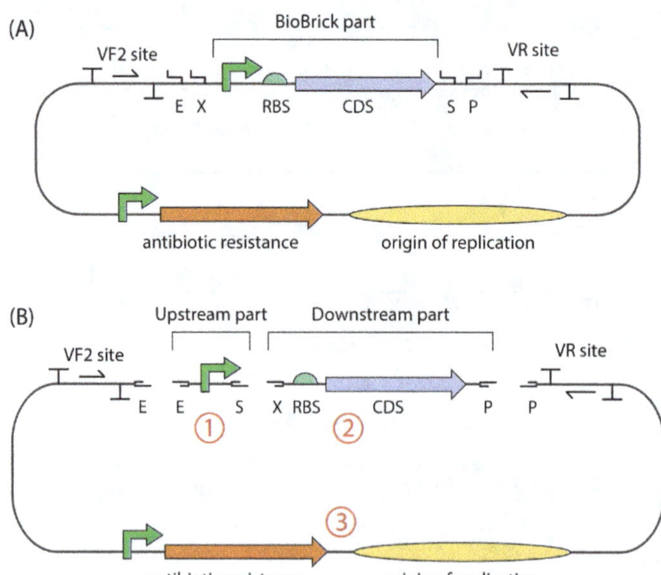

Fig. 24   (**A**) Illustration of a plasmid containing a promoter, RBS and coding sequence (CDS) of interest. Primer sites for later colony PCR are also shown. (**B**) Construction of the plasmid in **A** from two BioBrick™ parts numbered 1 and 2 (according to the scheme in Fig. 23B) and a destination plasmid vector labeled 3. E, X, S and P represent EcoRI, XbaI, SpeI and PstI restriction sites, respectively.

that this plasmid also contains downstream (between the S and P sites) a reporter CDS for a red fluorescent protein (RFP, i.e. mRFP1 codon optimized for *E. coli*; Shetty *et al.*, 2011); this is not shown in Fig. 25 because this CDS can be neglected using the standard 3A assembly method.

2. "Mystery" plasmid containing an RBS-chromoprotein CDS insert BBa_ K10339## bounded by BioBrick™ restriction sites (downstream part; Table 1 in the Appendices).

3. Medium-low copy plasmid pSB3K3 containing the RFP reporter gene insert bounded by BioBrick™ restriction sites (destination vector or backbone part). **A parallel cloning is also recommended** using a different destination vector, high copy plasmid pSB1K3, as comparing the color intensities produced from the two different vectors will aid in identification of the mystery chromoprotein gene.

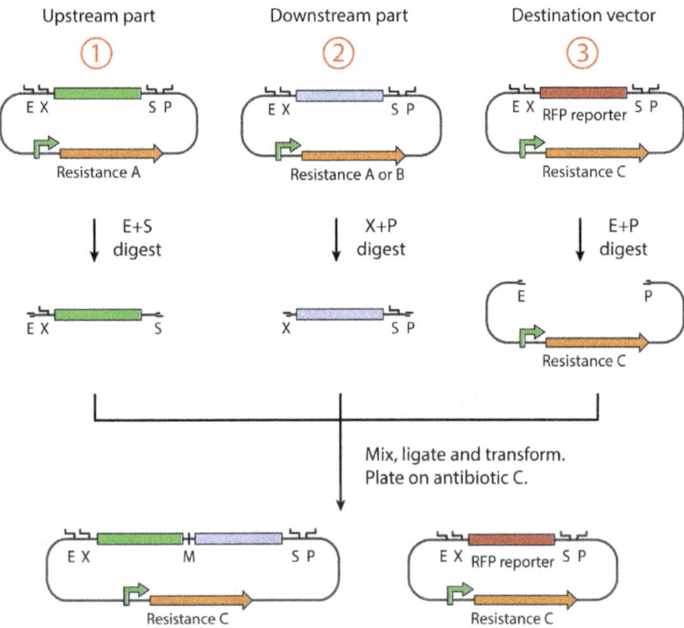

Fig. 25   A type of BioBrick™ assembly called "3A assembly" (Shetty *et al.*, 2011). Although 3A is an abbreviation for 3 Antibiotics, only two different antibiotic resistances are required. This is a second-generation BioBrick™ assembly designed to circumvent gel purification of restriction fragments. Here, the upstream part and the downstream part are digested out of their respective plasmids and ligated into a new (destination) vector plasmid encoding a different antibiotic resistance. This resistance allows selection against re-ligations among fragments just from the two plasmids containing the upstream and downstream parts. Selection against re-ligation of cut destination vector parts is not possible based on antibiotic resistance. Such unwanted events give red colonies if the RFP reporter fragment is included and white colonies for empty vector dimer (preventable by dephosphorylation of the vector). The numbering and color schemes are those of Fig. 24. M, mixed site or scar. Part 1 for our lab course contains an RFP reporter CDS between the S and P sites that is not shown here; this reporter can be neglected when using the standard 3A assembly method.

Plasmid copy numbers available in the Registry

The plasmids used in this manual have either the medium-low copy origin P15A (pSB3X#; copy number claimed to be 10–12) or the high copy origin pMB1 (pSB1X#; copy number of 500–700), and they are compatible with each other. When we set up this course, there did not seem to be any functional low copy plasmids available from the Registry of Standard Biological Parts. Although it contained plasmids based on the low copy origin pSC101 (pSB4X#), these have displayed anomalously high copy behavior in our university and in the hands of other iGEM teams.

The work flow is shown in Fig. 26.

Note that the promoter strain 1 has red colonies due to its RFP reporter, each mystery strain 2 has white colonies because it does not express its chromoprotein CDS, and the two alternative destination vector strains 3 have red colonies due to expression of the RFP reporter.

Protocol 2 (Chapter 6) is next, amplifying and purifying plasmids. Make up and inoculate four media solutions, each containing the appropriate antibiotic (Table 1 in Appendices) corresponding to the antibiotic resistance encoded by each of the four plasmids.

Antibiotic selection markers

When transforming and growing bacteria, it is important to select for only those cells that carry the desired plasmid. This is accomplished by incorporating an antibiotic resistance gene into the plasmid backbone and by growing in the presence of the antibiotic; only cells carrying the antibiotic resistance gene will be protected and survive (positive selection). Resistance genes can work through different mechanisms, such as pumping the antibiotic out from the cells or enzymatic degradation of the antibiotic. Note that some antibiotic-degrading enzymes are secreted from the cells, meaning that with time, even non-resistant cells can grow. This is mainly problematic with **ampicillin resistance** and can be observed on a plate when larger colonies of resistant bacteria become surrounded by smaller colonies of sensitive bacteria.

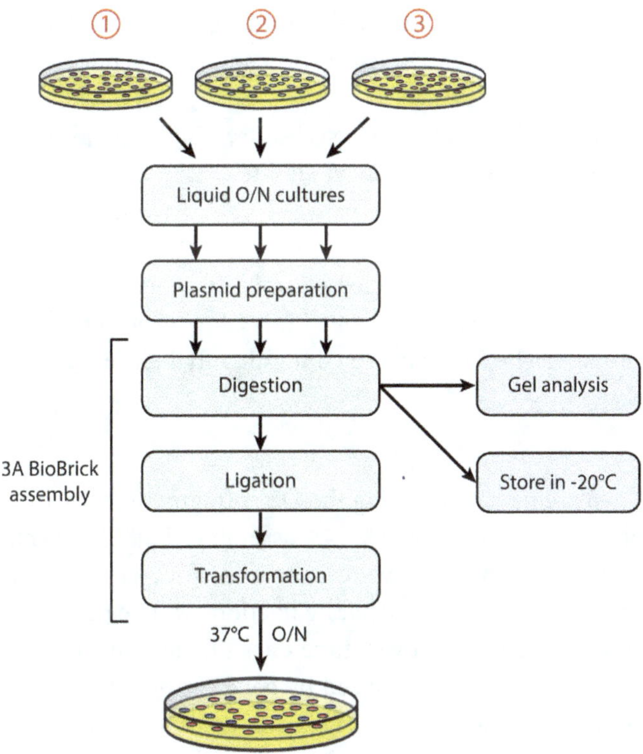

Fig. 26   Work flow for 3A assembly starting from three agar plates corresponding to parts 1, 2 and 3 of Fig. 25. For the destination vector part 3, trying two alternatives with different copy numbers is recommended.

By the next morning, all four cultures should have become very cloudy, i.e. grown to saturation. Now the four plasmid DNAs can be harvested using a plasmid prep kit according to the manufacturer's instructions. All four plasmid preps should give column eluates with high optical densities at 260 nm. Can you attribute any differences between yields to intrinsic features of the plasmids or procedural variables?

Plasmid miniprep kits

Kits from different manufacturers have different components and procedures. Therefore, always follow the manufacturer's instructions precisely. The first step is overnight culturing in 1–5 mL of medium. The cells are collected by centrifugation and resuspended in a buffer containing ribonuclease.

Next comes the addition of lysis buffer, a highly alkaline solution of the detergent, sodium dodecyl sulfate (SDS). The lysate is then neutralized and centrifuged to pellet cell debris, which includes cell wall, denatured proteins and chromosomal DNA. The supernatant, which is enriched for supercoiled plasmid DNA, is then loaded onto a silica column in the presence of a high salt concentration. The column is then washed to remove contaminants before elution of the plasmid.

Protocols 3–6 are next, generating the DNA fragments shown in Fig. 24B and assembling them by BioBrick™ 3A assembly (Fig. 25). Assuming two different destination vectors were prepared, four separate restriction digestion incubations and two separate ligations will be performed.

Results will be analyzed over three days (Table 7 in the Appendices). On the first day, digests are analyzed by gel electrophoresis. Then, in the morning of the third day, the cloning is analyzed by visualization of colonies on plates. Colonies harboring plasmids with a promoter upstream of a chromoprotein gene will have the color dictated by that particular chromoprotein. Does your choice of colored paper placed beneath the agar plate affect your sensitivity in detecting faintly colored colonies? How would a blue light with orange glasses (Safe Imager, Invitrogen;

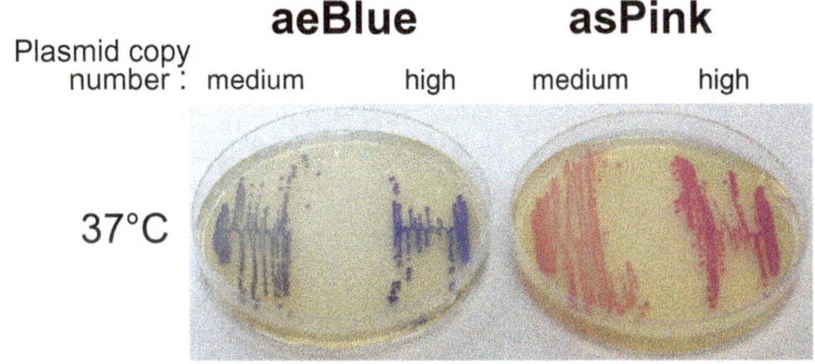

Fig. 27 Effect of plasmid copy number on color intensity at 20 hr incubation. Comparisons are best performed on the same plate. Taken from Liljeruhm *et al.* (2018) with permission.

Fig. 36 below) or a UV lamp help? Can you tell which chromoprotein gene you have? Note that chromoprotein color development may be slow due to the need for chemical maturation (Chapter 2). If some colonies on the plates are white after overnight incubation at 37°C, incubating them at 42°C for a couple of hours in the morning may speed up color development. Although this cloning experiment is not designed to generate white colonies, they usually occur anyway: what might they be? Document the results of each test and control plate: detailed descriptions may be sufficient, but cell-phone photography is encouraged (Fig. 27).

Protocol 7 is next, amplifying and characterizing promising plasmids. This involves amplification/purification by re-streaking (Figs. 27 and 28), liquid culture, plasmid preparation and characterization by digestion/gel electrophoresis and DNA sequencing (Fig. 29).

The analytical digestion is best performed by excising the insert with EcoRI and PstI, the two enzymes used to cut the destination vector. This will test if the insert is the size expected for giving the bacterial color observed (although it does not rule out multimeric inserts or vectors) and if the ends of the insert and vector are still intact (ends are susceptible to exonuclease "nibbling" of one or a few nucleotides during cloning). Finally, if the digest gave the correct-sized fragments, the plasmid should be sent for DNA sequencing across the insert using the VF2 primer (Fig. 24A). The exact method of preparation of samples for sequencing varies according to the sequencing facility used.

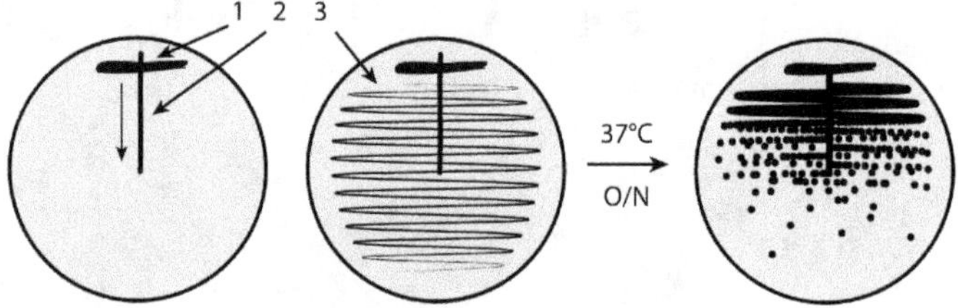

Fig. 28   Generating many single colonies by re-streaking out one colony. This three-step procedure is the alternative 1 detailed in Protocol 7.

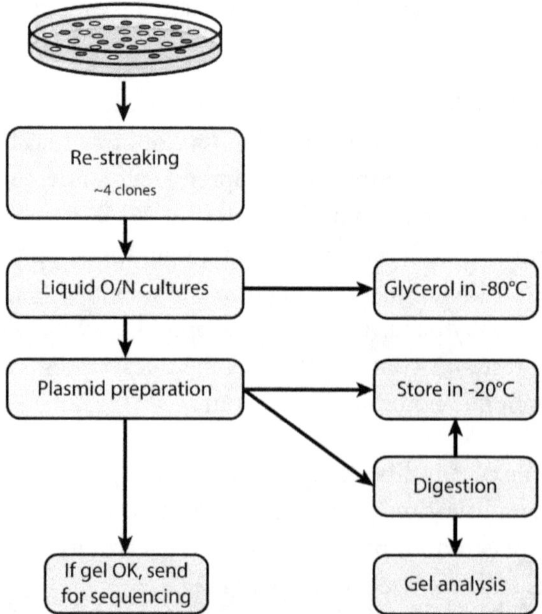

Fig. 29   Work flow for characterizing the colonies of a 3A-assembly transformation plate.

Sequencing DNA

Upon submitting plasmid or PCR DNAs to a sequencing facility, most commonly, Sanger sequencing is performed. In Sanger sequencing, a synthetic oligodeoxyribonucleotide primer is hybridized upstream of the region of interest and the primer extended on the template by a DNA polymerase.

Besides the four dNTPs, four dideoxyNTPs with specific fluorophores are also added to the reaction. The dideoxyNTP terminates the primer extension due to the lack of a 3′OH group necessary for phosphodiester bond formation. The products of the primer extension are then separated through a column and, at the end, a laser excites the fluorescent label on the last nucleotide. A camera detects the color and this reveals the sequence of the template, one nucleotide at a time. Beyond Sanger sequencing, next-generation sequencing technologies enable much higher throughput and can be much more time and cost effective, depending on the type of project.

While waiting for DNA sequences, students may either: (i) finish Lab section 2 by performing colorimetric assays (see next paragraph) or (ii) get a head start on Lab section 3 (see below) by designing and ordering PCR primers because primer synthesis and delivery takes a few days (Table 7 in Appendices).

Colorimetric Assays for Chromoprotein Expression

It is now time to practice an important element of synthetic biology — measurement of the activities of DNA parts. This assay is necessary to determine the effects of mutations that you will create for up- and down-regulating chromoprotein expression. Quantitating this is more difficult than you might expect, so there is still no accepted standard method and creativity is encouraged! Visual inspection of the colors of the bacteria in solid and liquid cultures and in pellets provides a qualitative estimate of chromoprotein expression level, and this can be documented readily by color photography. For the fairest comparison, re-streak high and medium-low copy clones on the same plate. But can you quantitate expression levels with numbers?

We used cameras and the programs **Photoshop** (proprietary) or **Image J** (free) for quantitation (Liljeruhm *et al.*, 2018), mindful that unfolded or unmatured chromoproteins are colorless and thus escape detection (Bao *et al.*, 2020). The software processes color photos by converting them into gray-scale images that separate each of the red, green and blue intensities for subsequent addition. However, we noticed that colonies on a plate that are darker red (by eye and by visual inspection of the photo), while increasing the red channel measurement as expected, can *decrease* the green channel measurement by a *greater* magnitude, leading to a three-color intensity addition that is actually lower! Hence, it may be prudent to focus on one color or wavelength.

- If you pick a wavelength(s), how would you decide?
- Might a spectrophotometer be helpful?
- If your chromoprotein absorbs light at OD_{600}, the wavelength typically used to monitor *E. coli* growth, how can you control for different extents of growth?
- What is your signal-to-noise ratio?
- If you have access to a flow cytometer, might that be helpful (see Figs. 40 and 41 in Chapter 7)?
- Might cell lysis be helpful in reducing background absorbance by the cells?

Protocol 8 is for lysis with lysozyme.

- Can you achieve lysis efficiently with this lysozyme protocol?
- How can the extent of cell lysis be estimated easily?
- Once you develop an assay for measuring expression level, how reproducible is it?
- What are your standard deviations and standard errors of measurement?
- What are your P values, and are they significant?

In preparation for these colorimetric assays, remember that more replicates will give better statistics. Also, **several types of overnight liquid cultures are advised:**

1. Your chromoprotein-expressing strain (test).
2. Its precursor strain lacking the promoter (control for the test; may not be a true negative control because of possible cryptic promoter activity upstream of the cloning site).
3. The cloning strain lacking a plasmid (negative control).

4. The most strongly colored strain from your lab class (positive control for lysis).

Ultimately, which of your quantitation methods seemed more reliable and why?

When your sequencing results come back, both the computer-deduced sequence and the raw sequence data will be provided. Software programs are available to help with the analysis. Are all the nucleotide peaks clearly interpretable or is your "clone" actually a mixture? Are the sequences of the parts correct? These sequences can be downloaded from the Registry of Standard Biological Parts (p. 43) by typing in the BBa-code for the plasmid name (e.g. BBa_K1033931 brings up the sequence of amilGFP; Table 1 in the Appendices). Are the junctions, such as the scar sequence, correct? If there is a mutation in the insert, would it be predicted to decrease the fitness cost of the plasmid to the cell?

Fitness costs and genetic stability

The function of synthetic systems in bacteria is always limited by the "metabolic budget" of the cells. Expression of foreign proteins or production of metabolites will consume both energy and resources that otherwise could have been used for growth, so the expression of synthetic systems almost always confers a fitness cost, making the bacteria grow slower than they otherwise would have done. When using strong promoters or when expressing toxic genes, this fitness cost can get very high, leading to a strong selection pressure to lose the genes responsible for this cost. If any bacteria in a population get loss-of-function mutations in any of the genes causing the fitness cost, they will have a growth advantage compared to the surrounding bacteria, and they will be rapidly enriched in the population. This is especially problematic when amplifying high copy plasmids in liquid media, since a plasmid present in 100 copies per cell also will have a 100-fold bigger chance to pick up loss-of-function mutations compared to a construct present in only one copy. Simple loss-of-function mutations are frame shifts and stop codons within coding sequences. Recombination between homologous sequences flanking an insert is favored if one part is reused several times in a construct, e.g. when using the same standard terminator both downstream and upstream of an insert. This is one of the reasons it is important for the synthetic biology community to create larger libraries of standard parts that lack homologies with each other.

Lab Section 3: Rational Engineering of Chromoprotein Expression Level

Synthetic biologists usually alter protein expression from a plasmid by changing:

1. the plasmid copy number, affecting the gene dosage;
2. the promoter, affecting initiation of transcription (Fig. 5A);
3. translation initiation sequences (Fig. 7);
4. the codon bias of the protein coding sequence (CDS), affecting translation initiation and elongation (Fig. 9).

Less commonly, synthetic biologists adjust gene expression by changing:

5. translation by targeting the mRNA with antisense RNA (Fig. 11);
6. the stability of the mRNA;
7. the stability of the protein.

Expression can be regulated by changing:

8. the RNA polymerase;
9. transcription by adding an operator sequence for binding a repressor protein that is inducible with a small molecule (Fig. 5B).

Change 1 has already been tested above by all groups, so now the groups will test each of Changes 2–5. Each group will pick one of these Changes 2–5 to pursue and attempt both up- and down-regulation of chromoprotein expression (except for Change 5, where up-regulation cannot easily be designed). If there are more than four groups, necessitating that two groups pursue the same category of change, they should test different hypotheses.

From this point on, the designing and testing of specific hypotheses is driven by the creativity of the students (like iGEM), rather than teacher-driven. So, rather than providing detailed experimental design, we provide a list of questions and additional general experimental protocols helpful for answering these questions experimentally. Any mutations up to 80 straight base pairs in length are compatible with the mutagenesis method provided. So is randomization of several bases to create libraries.

Changes 2–5

- Using the **background given in Chapter 2** and any other information available in textbooks, scientific papers or on the Internet, can you design specific mutations to effect your desired changes in expression?
- First, contemplate the plasmid you start with. If you made both high and medium-low copy number plasmids expressing your chromoprotein, are one or both suitable for testing up- or down-regulation, or would it be better to start with a plasmid(s) of another lab group? We picked two of the chromoprotein genes partly because they had not been codon optimized, thus making them ideal for codon bias experiments.

Change 2, promoter

- How strong is your promoter sequence?
- How does it compare with other promoters in the Registry of Standard Biological Parts and the maximal promoters CP25 and apFAB46 (Liljeruhm *et al.*, 2018)?
- What is the relative importance of the −35 box, the −10 box and the spacing between them?
- Can you regulate plasmid transcription by inserting just a transcription enhancer sequence or a repressor's operator sequence? Consider the relative dosage of the cognate chromosomally encoded enhancer/repressor protein.

Change 3, translation initiation

- How strong is your RBS sequence and how should you change it (Fig. 7A)?

- Is your RBS likely inhibited by base pairing with adjacent sequences in the mRNA (Fig. 7B)? Computational secondary structural modeling is useful (e.g. **mfold** Web Server).
- Can you change translation initiation by mutating sequences adjacent to the RBS in the mRNA?
- Can you activate translation with an epsilon translation enhancer sequence (e.g. Forster *et al.*, 2001)?
- Should the translation start codon be mutated?

Change 4, codon bias

- Has the CDS of your chromoprotein been codon optimized for *E. coli* (Table 1 in the Appendices)? If not, it is better suited for codon bias experiments.
- Is the codon usage significantly different from that of highly expressed genes in *E. coli* (Fig. 9)?
- Are there any codons at the start of the CDS that might slow ribosome clearance from the initiation region or affect base pairing with the RBS?
- How big is the footprint of the ribosome on the mRNA?
- Are there rare codons further downstream?
- Should all the codons in the region you change be substituted with the most optimal codons?

Change 5, antisense

- What region of the mRNA would you target with antisense RNA?
- How long would you make the antisense sequence? (See Vogel *et al.*, 2020).
- How might this affect toxicity?
- What are the pros and cons of encoding the antisense RNA on the same plasmid versus a separate one?
- If a separate plasmid is used for the antisense RNA, is it compatible with the chromoprotein plasmid? For more information and examples of antisense results, see Fig. 35.

Protocols 9 and 10. The work flow for PCR mutagenesis and analysis (Fig. 30) differs from previous work flows in that it incorporates PCR (Fig. 31).

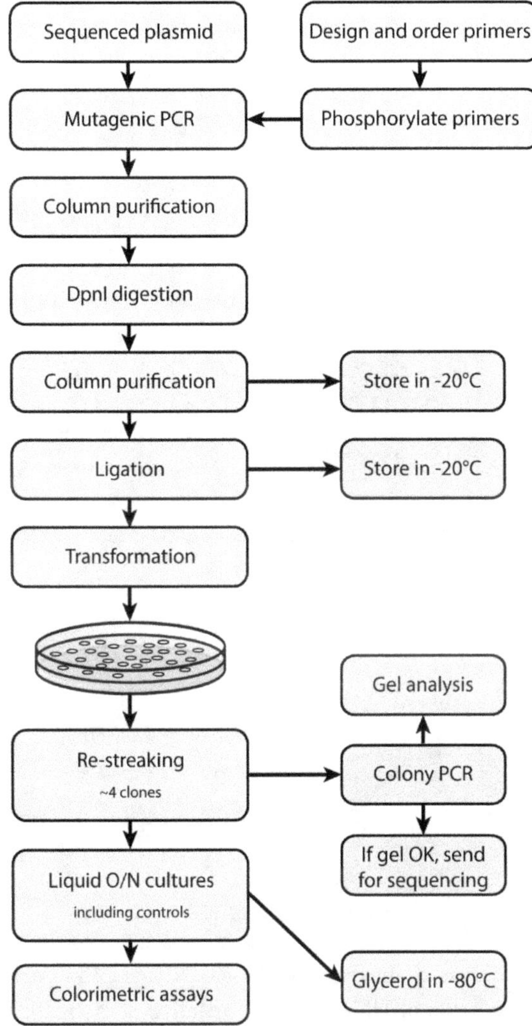

Fig. 30 Work flow for PCR mutagenesis and analysis.

The version of PCR we will use now is shown in Figs. 32 and 33A.

After the PCR reaction, the remaining plasmid DNA is degraded by DpnI, a restriction endonuclease that cuts a specific methylated DNA sequence (Fig. 33B).

If the transformation platings from PCR mutagenesis look promising, proceed to re-streaking and **Protocol 11**, colony PCR, to amplify your DNA for sequencing.

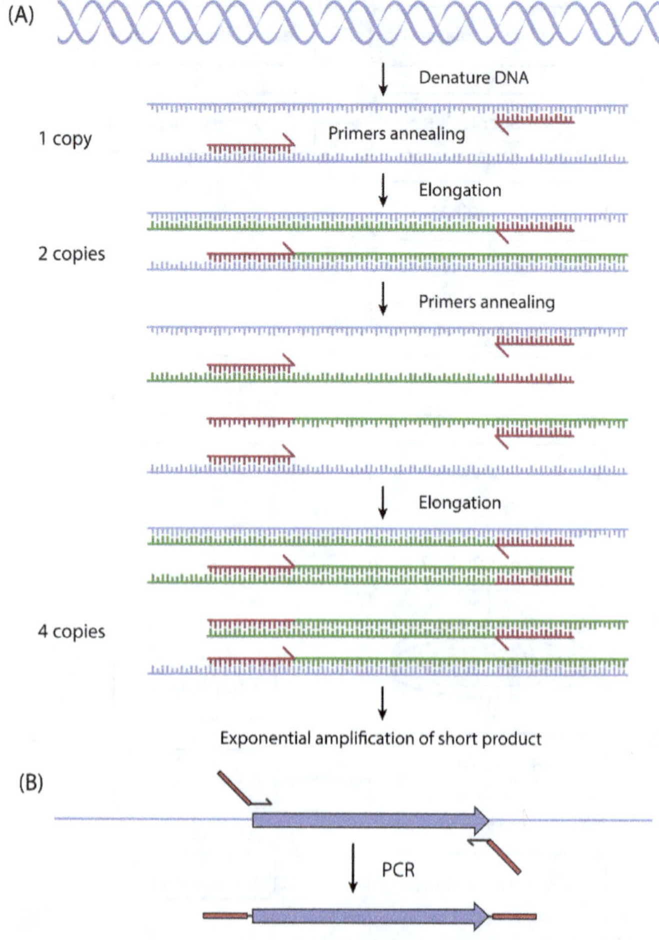

Fig. 31 The polymerase chain reaction (PCR). PCR is an in vitro technique for amplifying short segments of DNA. (**A**) The reaction starts with melting at high temperature of the double-stranded template DNA (blue) into single-stranded DNA. This is followed by a lower temperature step where two synthetic oligo primers (red) can bind specifically (anneal) to their target sequences. The third step is elongation, where a thermostable DNA polymerase synthesizes new strands of DNA (green) using the original template strands as templates. The DNA is then melted again and the cycle repeats. For every cycle, the amount of the short product is doubled, leading to an exponential growth of product. (**B**) PCR can be used to modify easily any DNA sequence. The extra bases at the 5' ends of the primers (single-stranded "overhangs" in red) are added to the ends of the final amplified sequence. Note that primers should be designed with extra bases 5' of any restriction sites to guarantee subsequent cutting (see New England Biolabs catalog).

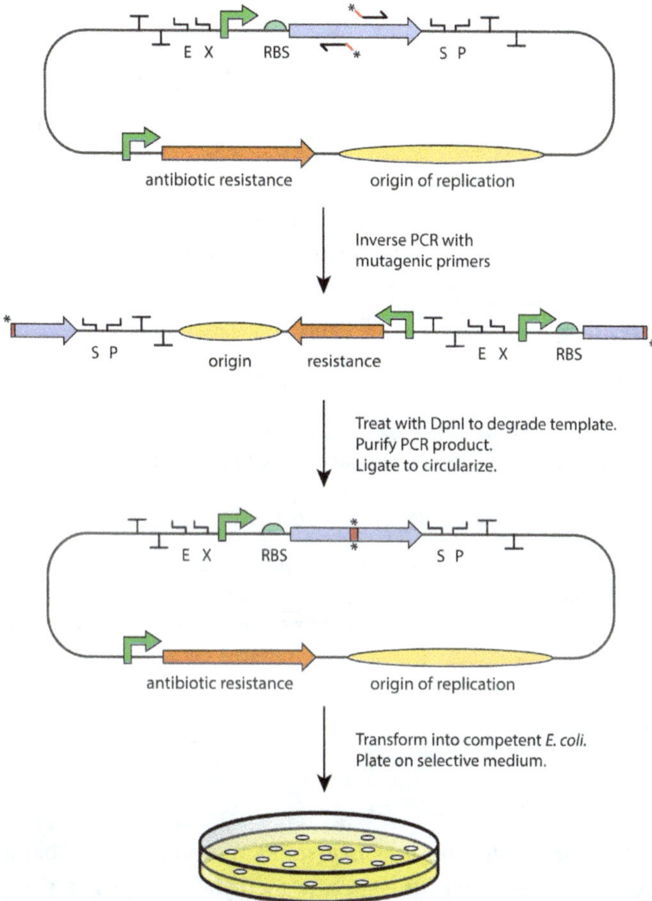

Fig. 32 Inverse PCR mutagenesis. The phosphates introduced by phosphorylation of the purchased synthetic oligodeoxyribonucleotides are marked with asterisks. The new sequences created (red) can be up to 80 straight bps and are derived from both primer overhangs.

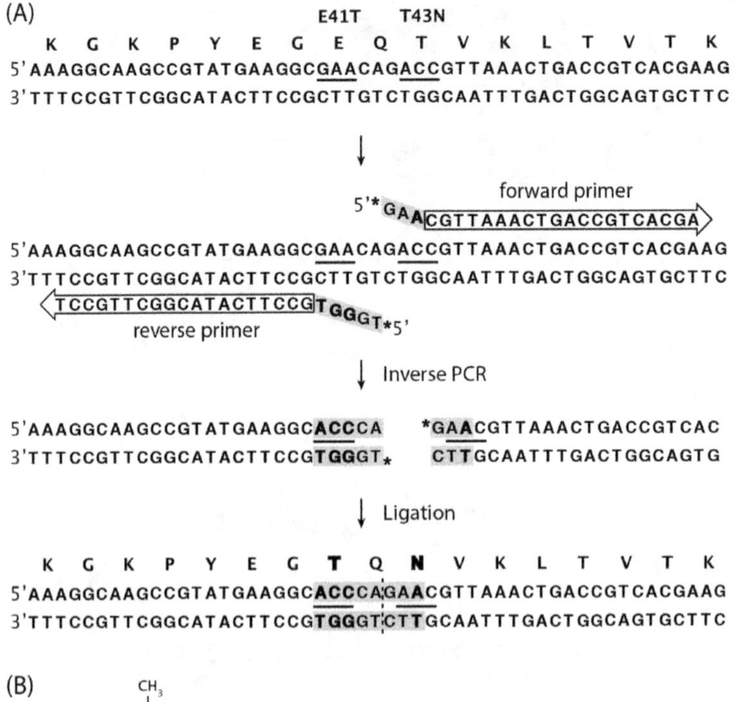

Fig. 33 (A) Primer design for inverse PCR mutagenesis changing two amino acids. (B) DpnI cleaves the specific methylated sequence shown at the sites indicated with triangles. The methylation only occurs in vivo, so plasmids are cleaved, whereas PCR products are unmethylated and thus resistant to cleavage.

While waiting for sequencing results, use your colorimetric assays to quantitate expression to test your hypotheses concerning changing expression levels. Alternatively, if any prior BioBrick™ cloning or PCR mutagenesis experiments failed, they can be repeated or trouble-shooted.

Suggestions for trouble-shooting failed experiments

- **Liquid culture of chromoprotein-expressing strain**
  - Color is absent or incorrect
    - Not enough oxygen supply
    - Contaminative growth
  - Color is weak or varies between preps
    - Not enough oxygen supply
    - Mutation(s) was selected mid way during growth due to high fitness cost of very high expression of chromoprotein
      - Decrease expression level or temperature

- **PCR**
  - Multiple different products or incorrectly sized product
    - Annealing temperature is too low. Note that the recommended annealing temperature is above or equal to Tm for Phusion® pol, but below Tm for Taq pol.
    - Mis-priming and extension occurred during initial denaturation step
      - Add the DNA polymerase to the hot wall of the tube after reaction mix reaches the denaturation temperature, then mix (without removal of the tube from the PCR machine)
    - Too many reaction cycles
    - DNA contaminants
      - Purer template needed
    - Primers are non-specific
      - Order new (longer?) primers
      - Extract the correct product from the gel to purify from other products
      - Order two more primers for nested PCR
      - If there are extra bands below ~1000 bp in PCR mutagenesis, continue anyway according to the procedure. The origin of replication is incomplete in these small products, so they will not replicate in vivo

- o Product is absent or yield is low
  - Too little template
  - Too much template
  - Denaturing time is too short
  - Annealing temperature is too high. Try gradient PCR if machine allows.
  - Too few reaction cycles
  - Re-check Tm calculations for the primers
  - Poor primer design
    - □ Order new primers or add DMSO
  - A component was left out of the PCR reaction
- **Ligation and transformation steps in PCR mutagenesis**
  - o Few or no colonies after transformation
    - Transformation efficiency is too low
      - □ Remake competent cells or electroporate (p. 131)
    - Transformation used too much DNA, which is toxic to the cells
    - Non-functional or incorrect plasmid
      - □ Prior steps failed
    - The new mutant is toxic to the cell
      - □ Use a medium-low copy number plasmid
    - Add PEG to ligation
  - o Too many colonies after transformation
    - The antibiotic in the plates has degraded
    - The DpnI treatment of the PCR template was inefficient
      - □ Sequence several colonies
      - □ Perform colony PCR screening if the mutation is expected to produce a difference in size measurable on an agarose gel
  - o Deleted nucleotides at the blunt-end ligation site
    - Phusion® pol has 5′ exonuclease activity if
      - □ incubation is too long
  - o Extra nucleotides at the blunt-end ligation site
    - Taq pol was used by mistake. This pol adds on an extra A to the 3′ end

Our five-week course provided sufficient time for a second round of PCR mutagenesis. This can be performed directly on bacterial colonies (Protocol 11), obviating another plasmid prep. Aims in this second mutagenesis round included:

- Second-generation mutational design.
- Alteration of cell color by placing two different chromoproteins in the same operon. Construction by 3A assembly requires another destination vector (pSB1A3, which is resistant to amp; Table 1 in the Appendices) with different antibiotic resistance to the upstream part (promoter-fused chromoprotein gene in a vector resistant to kan) and downstream part (promoterless chromoprotein gene in a vector resistant to chloramp). Alternatively, try Gibson assembly (see below and Protocol 12).
- Alteration of the color of the chromoprotein itself (Fig. 34).
- Regulation of expression. This can be done by subcloning into a pET vector (see p. 13). Alternatively, a Lac operator can be inserted, and the plasmid transformed into the DH5α LacIq chassis.

Another example of class experimental results is shown in Fig. 35.

If simpler protocols are preferred, e.g. **when lab time is limited to as few as two days, or for high school students**, subsets of the protocols suffice. Chromoprotein identification in Lab Section 2 can be readily streamlined by substituting BioBrick™ cloning steps with a pre-constructed plasmid and pre-made competent cells for transformation. An even simpler version is supplying a mixture of our colored bacteria for separation and identification by streaking on a plate (Fig. 36).

Lab Section 4. Other Methods and Experiments

Building genes and operons is easiest by the BioBrick™ method, but it has limitations. The scar sequence is unavoidable and reduces flexibility, e.g. when joining a Shine–Dalgarno sequence with a coding region or when fusing two coding regions. The former can be addressed with a shorter prefix (see Fig. 23A legend), while the latter can be addressed with alternate BioBrick™ assembly standards that do not shift the reading frame. Another limitation is that most source DNAs lack convenient restriction

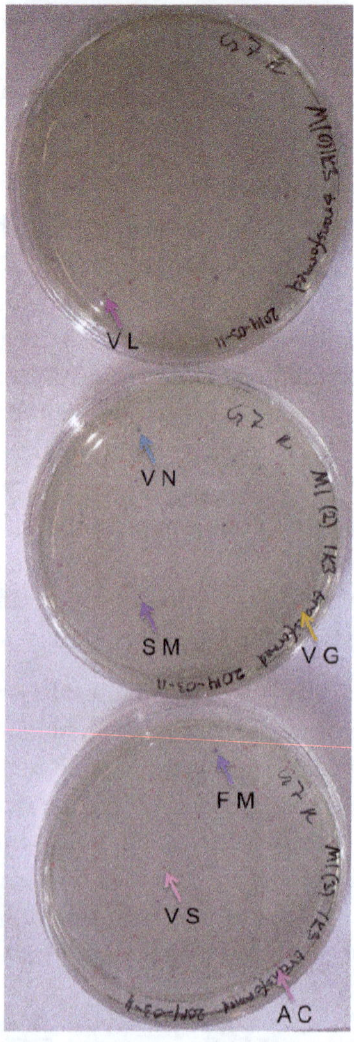

Fig. 34 Color-directed mutagenesis of a chromoprotein. To our knowledge, this experiment by our lab course students produced the highest frequency of chromoprotein color changes ever reported! Positions C64 and Q65 of amilCP (GFP numbering) were mutagenized randomly ($=20^2$ possible dipeptide sequences), transformed into *E. coli* DH5α and plated on LB kan. We counted 518 colonies in four main classes: amilCP-like intense purple-blue (26), weak purple-blue (53), white (389) and new color (50). An assortment of new colors was picked (arrowed) and amplified to produce Fig. 4A of Liljeruhm *et al.* (2018). Sequences of the mutated amino acid pairs are given (see Fig. 9 for single-letter abbreviations).

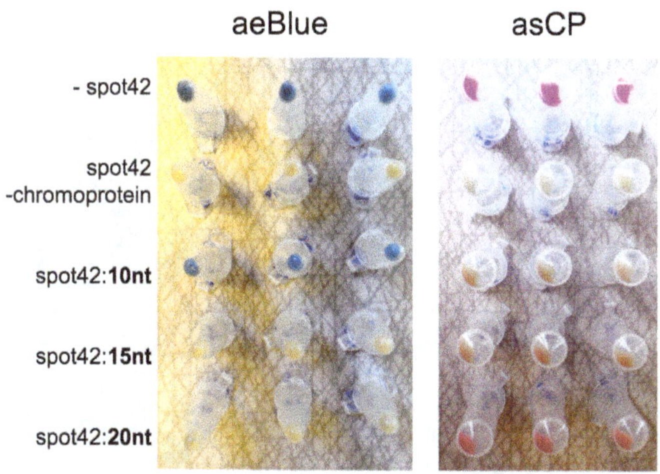

Fig. 35 Fine tuning of chromoprotein expression using sRNAs. Sequences complementary to the very first few codons of chromoproteins aeBlue and asCP (i.e. 10, 15 and 20 nts complementary to AUG…) were inserted by inverse PCR mutagenesis into the artificial sRNA based on spot42 in vector pSB1C3 (BBa_ K864440 in Table 1 in the Appendices; see also Figs. 11A and 50) at the site shown as a randomized region in Fig. 1b right of Sharma *et al.* (2011). Cells transformed with the combinations of chromoprotein-expressing plasmids (in vector pSB3K3) and spot42 plasmids indicated in the figure were amplified and pelleted in triplicate to give the results shown. The variation in inhibition of chromoprotein expression with length of complementary sequence differed significantly for the two different chromoproteins. Experiment performed by Uppsala University students P. Enström, A. Gynnå, M. Rahmani and N. Sandberg. (See Vogel *et al.*, 2020.)

sites for cloning. Although flanking BioBrick™ sites can be introduced readily by PCR (Fig. 31B), this is insufficient to convert a gene into a BioBrick™ if there is an "illegal" BioBrick™ site(s) internally. Fortunately, there are several solutions. Commercial gene synthesis enables illegal BioBrick™ sites to be changed readily to synonymous codon sequences. We showed scalability up to assembly of 30 BioBrick™ cistrons totaling 58 kbp (Shepherd *et al.*, 2017). Multifragment mutagenesis is feasible (Wäneskog and Bjerling, 2014) using overlap extension PCR (Fig. 37). Alternatively, there is Gibson assembly.

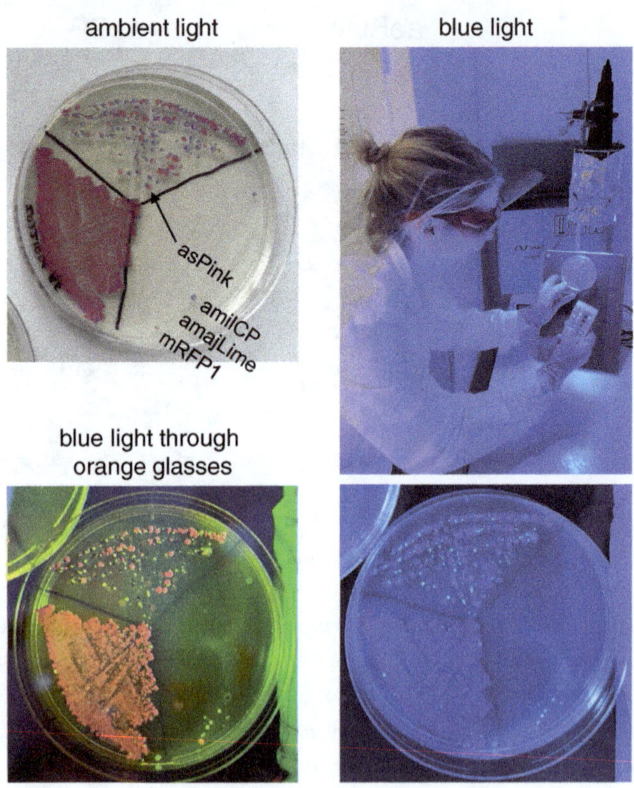

Fig. 36   Identification of chromoprotein plasmids in a bacterial mixture. Aliquots of separate overnight cultures of *E. coli* MG1655 containing high copy plasmids expressing amajLime (200 µL as plasmid is quite toxic), amilCP (50 µL), asPink (10 µL) and mRFP1 (10 µL) were mixed, streaked on a chloramp plate and incubated at 37°C for 18 hr. The plate was photographed in the three ways indicated. The orange glasses (Safe Imager, Invitrogen) were our best filter option to distinguish asPink from mRFP1. These four plasmids can all be derived from Addgene's kit 2.0 (Table 1 in Appendices) if asPink and mRFP1 are subcloned into a chloramp vector.

Protocol 12, Gibson assembly, circumvents any potential problems due to scars or unavailable/inconvenient restriction sites because it creates sticky ends without using restriction enzymes (Fig. 38). Also, unlike standard BioBrick™ assembly, it enables assembly into a vector of more

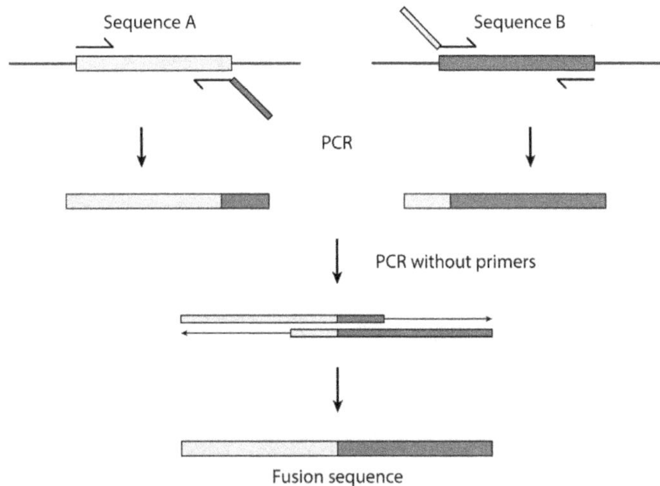

Fig. 37 Overlap extension PCR joins multiple fragments without using restriction enzymes. It is useful for multifragment mutagenesis and as an alternative to Gibson assembly.

than two parts at a time. Thus, this core technology of synthetic biology is often the method of choice for fusing DNAs together.

The main drawback of Gibson assembly is that it requires PCR, which has a much higher mutation rate than DNA amplification in vivo, especially with sequences containing repeats. Also, annealing of some ends can be inefficient, perhaps due to secondary structure formation at 50°C. This latter problem can usually be circumvented by overlap extension PCR (see Protocol 12 and Fig. 37). Alternatively, another popular, scar-free method for assembling several parts at once is cloning with type IIS restriction enzymes.

Unlike BioBrick™ restriction enzymes, type IIS ones cut outside their DNA recognition sites (Fig. 39), enabling the free choice of overhang sequence. Thus, by flanking a DNA part with two suitably oriented type IIS sites, it is possible to cut the part out of its vector and ligate it to another similarly removed part without introducing a scar. The main drawbacks of this approach are that illegal type IIS sites must be removed, and parts need to be assigned to pre-determined orders of assembly and levels of vectors. This complexity has led to the development of multiple standards including NOMAD, Golden Gate, GoldenBraid 2.0, MOBIUS, MoClo,

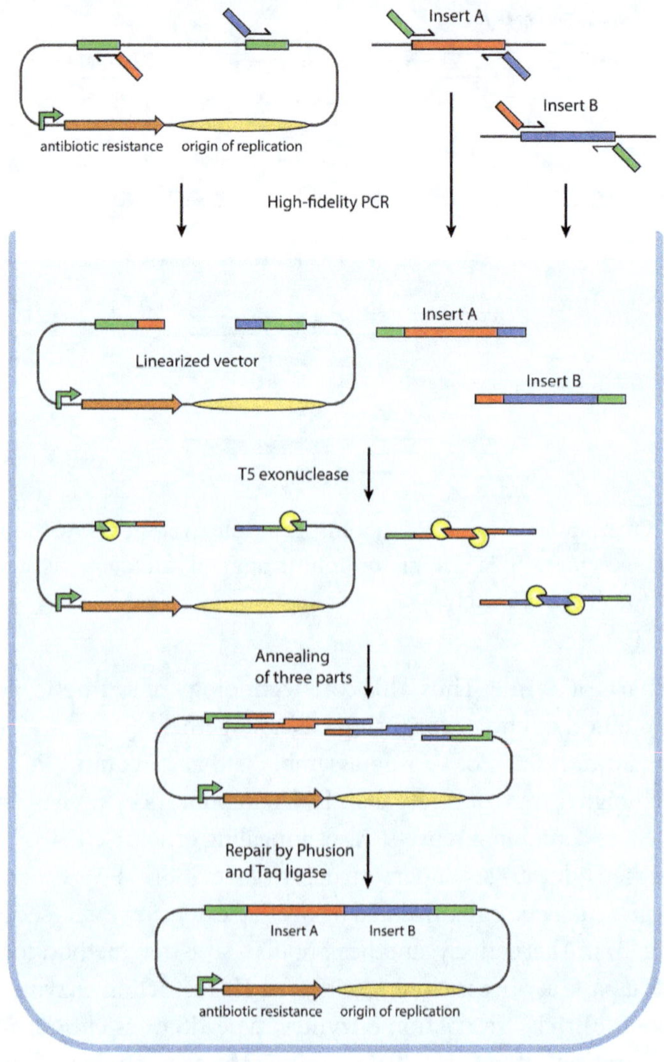

Fig. 38 Gibson assembly. This technique ligates one or more PCR fragments to a vector in one reaction mixture without restriction enzymes. First, PCR fragments are generated with primers that add overlapping sequences. Second, T5 exonuclease digests the 5' ends, creating sticky ends that anneal. Third, Phusion® HF DNA polymerase repairs the single-stranded regions. Last, Taq DNA ligase forms phosphodiester bonds between the DNA sequences to seal the plasmid.

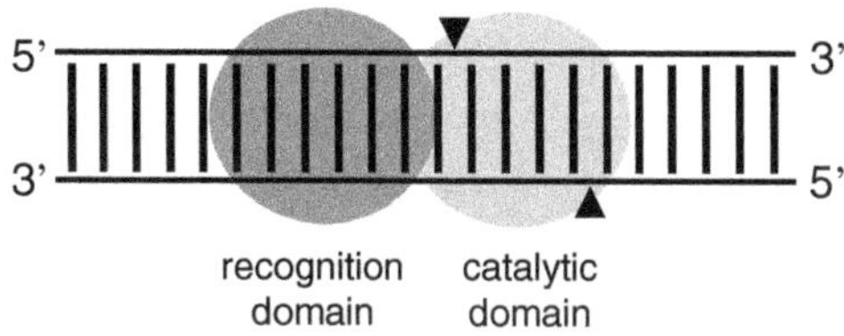

Fig. 39   Type IIS restriction endonuclease schematic. These enzymes cut (triangles) outside their DNA recognition sites, enabling free choice of overhang sequence.

LOOP and GoldBricks (Chaudhari and Hanson, 2021). Hence, all things considered, we still prefer BioBricks™.

Another important and flexible technology for joining DNA parts in synthetic biology is **recombination** in vivo. Like Gibson assembly, recombination methods such as Lambda Red recombineering also operate without restriction endonucleases. And recombination is an essential technique if genes are to be knocked out or inserted into the chromosome. Recombination is covered in Chapter 7 because it is deemed more suitable for research and iGEM than for initial teaching labs.

Computational Modeling Lab

Computer-aided mathematical design, analysis and modeling are increasingly important in SynBio as we better understand and quantitate the molecular systems and strive for ever-more-complex syntheses. Thus, our lab course was soon partnered with an inspiring "dry lab" created by Dr. David Fange. In his three-hour tutorial, students individually computationally model a simple genetic toggle switch.

Two inducible repressor genes (see Fig. 5B) that mutually inhibit each other were physically encoded on a plasmid in *E. coli* in the classic toggle switch of Gardner *et al.* (2000). One of the repressors was linked to a GFP reporter CDS whose activity (i.e. fluorescence) could be flipped between stable on and off states by transient induction. Gardner *et al.* modeled the bistability of their cells with the proprietary Matlab software, while Dr. Fange used the free COmplex PAthway SImulator (COPASI) software (Hoops *et al.*, 2006; www.copasi.org).

BioBuilder and Logic Gates

BioBuilder is a short practical course in synthetic biology for high school or early university levels. It is based upon prize-winning iGEM projects and is commercially available via kits (http://www.biobuilder.org) and a book (Kuldell *et al.*, 2015). Detailed information is provided for teachers on how to set up the labs and computer simulation, and detailed protocols are provided for the students.

The labs illustrate two of the logic gates increasingly used in SynBio. The logic gate term comes from an electronic circuit where multiple inputs are assessed together to control one output. Such gates form the basis of all digital devices. Importantly, the SynBio logic gates are modular, allowing them to be combined reliably to produce more complex behaviors. Examples of gates are:

AND, OR, NOT (Inverter), NAND (NOT AND), NOR (NOT OR).

We summarize the practicals briefly as follows:

1. Eau that smell

This lab is based on MIT's 2006 iGEM project. It uses banana odor as a readout of the growth phase of *E. coli*.

Three strains of *E. coli* that have different expression patterns of ATF1 protein are provided. This enzyme converts isoamyl alcohol to isoamyl acetate, the product responsible for banana's distinctive smell. Each strain expresses ATF1 from a different plasmid. One plasmid has a promoter that expresses the enzyme at stationary phase (BBa_J45250). The second plasmid has a different promoter (BBa_J45200), and the third plasmid has a genetic **inverter (NOT gate**; BBa_J45990), both of which switch expression to log phase. Students measure ODs and smell intensities from the three strains through lag, log and stationary phases.

2. iTune device

This lab uses β-galactosidase enzyme activity as a readout of expression levels of nine different strains of *E. coli*. The nine strains in the provided kit each harbor a plasmid encoding β-galactosidase downstream of a different combination of weak, medium and strong promoters with weak, medium and strong RBSs.

3. Picture this

This computer simulation is based on the 2004 iGEM project at the University of Texas, Austin. Different regulatory parameters in a signal transduction pathway activated by light are modeled.

4. What a colorful world

This lab is based on the University of Cambridge's 2009 iGEM project. Two different plasmid-encoded operons for color production are transformed into two different strains of *E. coli* to compare the effect of chassis on expression. The operons encode enzyme pathways that make small molecule pigments (BBa_K274002 and BBa_K274003). An **AND gate** enables different colors to be produced. The lab includes making competent cells through a simple version of the $CaCl_2$ technique.

5. Golden bread

This lab uses a eukaryote, a baker's yeast strain (*Saccharomyces cerevisiae*), which was genetically modified with a fungal pathway to produce β-carotene (an orange vitamin A precursor). It is based on Johns Hopkins University's 2011 iGEM project that revealed instability of the coloration during propagation. The lab involves transformation with the crtYB gene from the pathway to test its potential role in instability.

The "Dreaded" Exam

Though everyone complains about exams, quality examinations are not only necessary to document mastery of a subject, they are also appreciated very much by students. But how best to examine the practical portion of a synthetic biology course? Whatever the assessment form takes, we believe such a big part of a course deserves a big part of the grades. But it can be hard to give individual grades for lab sections, especially when students work in groups. What worked well for us, at least according to anonymous course evaluations from the students, was the following:

- The student's degree of involvement in the lab was graded.
- Lab books were submitted to teachers for feedback after one week, then re-submitted for grading after completion of the experiments. For

guidelines on lab book preparation, see "The Lab Notebook" earlier in this chapter. A potential alternative suggestion would be to write up a report on the lab in the format of a scientific paper.

- Each group gave an oral Powerpoint presentation where every group member had to participate equally.

The oral presentations were given to the whole class so all groups could learn about all of the experiments performed and everyone had the opportunity to ask questions. Presentation day was a major highlight of the whole synthetic biology course.

Presenting lab work orally using slides

- The temporal structure of the presentation should be:
  - Title slide. Include group member names
  - Outline slide
  - Introduction
  - Aims/hypotheses
  - Methods. Methods that were identical for all groups should only be mentioned briefly
  - Results
  - Conclusions/summary slide. Inconclusive results merit plenty of discussion, including potential future work
- Keep to the time limit (20 min plus 5 min for questions worked well). The duration should be checked in a rehearsal.
- About one slide per minute is a good rule of thumb.
- Less important material should be placed at the end of the presentation so that, if the group is running out of time, it can be omitted without destroying the take-home messages of the talk.
- All slides are to be on a white background with dark (black or dark colored) text using Arial or Helvetica fonts.
- Every slide should have a title describing its contents.
- One, maybe two, figures on each slide. Gels and their plots should be aligned.

- Gel lanes should be labeled with their contents, not 1, 2, 3…
- Graphs and bar plots are superior to tables. They should have clearly labeled axes, and the axes should be mentioned when discussing the graph. Include error bars and P values.
- Every slide should have some bulleted text, but this should be kept to a minimum.
- The presentation should not be memorized word for word or read from notes.
- Speakers should be engaged with the audience and should encourage pertinent interruptions, as long as they are not too long!
- Questions should be answered briefly to leave time for other questions. It is OK to say that you do not know!

6

Protocols

Introduction

Protocols are to be performed in conjunction with Lab Sections 1–3 (Chapter 5) and contain all the detailed methods (together with a couple of manufacturer's protocols) necessary for students to complete these three sections. The text at the beginning of some protocols is written with both students and teachers in mind. The calculations for some of the solutions are left for the students because students preferred this more realistic training experience.

Before beginning lab work, know the safety procedures concerning fires, chemicals, biologicals and equipment detailed in Chapter 4. Prior to and after work each day, clean your lab bench with ethanol to try to ensure a sterile environment.

Read through the entire protocol before starting the experiment to prevent delays and to plan on minimizing the time that key reagents are removed from storage.

When removing enzymes from storage, always keep them chilled on ice and not for longer than necessary, otherwise they may lose activity. Remember to always add the enzyme last to a fully mixed solution, then mix again thoroughly. This is because enzymes are sensitive to inactivation by pH and ionic conditions that deviate from their storage and reaction buffers.

Storage of biologicals at low temperature

Storage at low temperature is crucial for the reuse over long periods of time of biologicals such as cells, protein enzymes and nucleic acids. But cells and protein enzymes, in particular, can be killed easily if frozen and/or stored in the incorrect manner. Rules differ depending on the individual biological, and there are many standard variables to consider, such as:

- temperatures of −80°C, −20°C or 4°C;
- water: dry (by lyophilization), wet or frozen;
- buffer: add glycerol and/or reducing agent, etc.;
- freezing rate: snap versus slow;
- aliquots: if only one cycle of freezing and thawing is recommended;
- history: has the sample been frozen before, and if so, how many times?

So it is important to read about storage in our protocols and in the manufacturer's instructions and/or to ask the teacher if you are unsure. In general, frequently used enzymes and reagents are stored in a −20°C freezer. Enzyme solutions in this category typically contain glycerol to prevent freezing.

However, if you wish to store samples for a longer period of time or they are less stable, a freezer with a temperature of −70°C to −80°C can be key. Examples of −80°C samples are competent cells, glycerol stocks of cultures, and enzymes from large purifications. When going into −80°C boxes, it is important not to thaw other samples; i.e. only open the freezer or the box briefly, and make sure the door/lid is closed properly.

Protocol 1. Preparation of Solutions and Agar Plates

Introduction

This protocol details the preparation of every solution necessary to perform the lab course. A pH meter is not required (except for Protocol 8, where the lab teachers should perform the steps with the pH meter). Final pHs should be verified with narrow-range pH paper. ddH_2O is either doubly distilled or of similar high purity. Solids should always be dissolved in a smaller volume of ddH_2O than the final calculated volume. For some solutions, autoclaving may be required to completely dissolve the solids (e.g. agar).

> **Useful equations**
>
> moles = mass/formula weight,   where mass (m) is in g
> $\Rightarrow m = mol \times FW$
> molarity = moles/volume,   where molarity is in
> $\Rightarrow mol = M \times L$ M and volume is in L
> 1% = 1 g/100 mL,   where % is assumed to be m/v or
>   w/v, unless v/v or m/m is stated
>
> DNA concentration,   where OD_{260} = optical
> $= OD_{260} \times 50$ ng/μL   density at 260 nm

Notes:

0.9% NaCl (Also Known as Isotonic Solution), 10 mL

Introduction

This solution is used for suspending cells to avoid early lysis by osmosis. Use for resuspension and/or washing the cell pellet if it will be divided up, stored for a few days, or analyzed spectrophotometrically in the absence of the colored LB. Some laboratories use phosphate buffered saline (PBS) instead.

Components

- NaCl;
- ddH_2O.

Procedure

1. Check that the balance is calibrated.
2. Calculate how much you need to weigh out of the compound for making a 10 mL solution.
3. Add the salt to a glass bottle and add water to dissolve.
4. Make up to the final volume of 10 mL.

Notes:

50% (v/v) Glycerol, 50 mL

Follow the autoclaving instructions from the lab teachers.

Introduction

This solution is to be used for making cell glycerol stocks of important bacterial strains.

Components

- Glycerol stock;
- ddH$_2$O.

Procedure

1. Check which percentage of glycerol is in the stock.
2. Calculate how much volume you need of glycerol and how much water you need to add to reach a final volume of 50 mL.
3. Measure the glycerol in a measuring cylinder.
4. Add to a glass bottle and add water to make a 50% glycerol solution.
5. Autoclave for 20 min.

Notes:

1 M $CaCl_2$

Follow the autoclaving instructions from the lab teachers.

Introduction

$CaCl_2$ is used in a less concentrated form for making competent *E. coli* cells. Upon exposure to $CaCl_2$, the cell wall becomes fragile and at 42°C *E. coli* takes up foreign plasmids efficiently.

Components

- $CaCl_2$;
- ddH_2O.

Procedure

1. Look up the molecular weight of $CaCl_2$.
2. Calculate how much $CaCl_2$ you need to weigh to give 10 mL of 1 M $CaCl_2$ solution.
3. Weigh out the powder and add it to 8 mL water.
4. Stir and shake until it dissolves.
5. Make up to 10 mL final volume.
6. Autoclave for 20 min.

Notes:

10x TBE Buffer

Na_2EDTA or Na_4EDTA can be used since their different effects on pH can be neglected for this buffer. The concentration in the gel and running buffer should always be 1x TBE.

Introduction

TBE stands for the three components Tris, boric acid and EDTA. EDTA chelates Mg^{2+}, both inhibiting enzymatic degradation of nucleic acid and giving sharper bands on gels. There is no risk of contaminative growth in 10x stock solution, but over time, the salt may precipitate, requiring the preparation of fresh buffer. An alternative and perhaps safer buffer is 10x TAE (see below).

Components

- Tris base;
- boric acid;
- EDTA;
- ddH_2O.

Procedure

1. Dissolve the following reagents in 400 mL of ddH_2O using a magnetic stirrer:
   Tris to a final concentration of 0.89 M; boric acid to a final concentration of 0.89 M.
2. Dissolve Na_2EDTA or Na_4EDTA to a final concentration of 25 mM.
3. Adjust the volume to 0.5 L. The pH will be ~8.5.
4. Store at room temperature.

1x TBE working solution

Dilute the stock solution by 10x with deionized water (dH_2O) in a new 1 L bottle.

Notes:

10x TAE Buffer

Introduction

TAE stands for the three components Tris, acetic acid and EDTA. This is an alternative, perhaps less toxic, electrophoresis buffer than 10x TBE.

Components

- Tris base;
- Glacial acetic acid;
- EDTA free acid;
- ddH_2O.

Procedure

1. Dissolve the following reagents in 800 mL of ddH_2O using a magnetic stirrer:
   48.4 g Tris base;
   2.92 g EDTA free acid, and
   10.9 g glacial acetic acid.

   Note: for safety when handling concentrated liquid acids, wear safety glasses and gloves, use a fume hood, and never add water to excess acid.
2. Adjust the volume to 1 L. The pH should be ~8.2.
   Calculate the concentration of your solution.
3. Store at room temperature.

1x TAE working solution

Dilute the stock solution by 10x with deionized water (dH_2O) in a new 1 L bottle.

Notes:

SOB Medium

It is unnecessary for every group to make up SOB media as only 50 mL is used in Protocol 5 and LB is an acceptable substitute. Make sure that the chemicals below, including 5 M NaOH, and the autoclave are available before starting. Follow the autoclave instructions from the lab teachers. The finished media should be autoclaved within a couple of hours to prevent contaminative growth.

Introduction

SOB medium, or Super Optimal Broth, is used for preparing chemically competent cells. This protocol is adapted from Ausubel *et al.* (1999).

Components

- Yeast extract;
- Bacto™ tryptone;
- NaCl;
- KCl;
- ddH$_2$O;
- 5 M NaOH.

Procedure

Add the following to a 1 L bottle:

1. 0.5% (w/v) yeast extract.
2. 2% (w/v) Bacto™ tryptone.
3. NaCl to a final concentration of 10 mM.
4. KCl to a final concentration of 2.5 mM.
5. Add ddH$_2$O up to 600 mL.
6. Adjust the pH to ~7.0 with 120 μL 5 M NaOH.
7. Autoclave for 20 min within 2 hr.
8. Store at room temperature.

Notes:

LB Medium

Depending on the length of time spent in lab, be prepared to make LB several times, either due to high usage or contamination. Make sure that the chemicals below, including 5 M NaOH, and the autoclave are available before starting. Follow the autoclave instructions from the lab teachers. The finished media should be autoclaved within a couple of hours to prevent contaminative growth.

Introduction

Lysogeny broth (LB) is one of the rich media for bacterial growth and is the standard choice for *E. coli*. LB, was developed by Bertani and later optimized by Luria during the 1950s. It subsequently acquired different names, including Luria broth, Luria–Bertani media and Lennox broth, some containing different salt concentrations.

Components

- NaCl;
- Bacto™ tryptone;
- yeast extract;
- ddH$_2$O;
- 5 M NaOH.

Procedure

Add the following to a 1 L bottle:

1. NaCl to a final concentration of 0.17 M.
2. 1% (w/v) Bacto™ tryptone.
3. 0.5% (w/v) yeast extract.
4. ddH$_2$O to 600 mL.
5. 100 μL of 5 M NaOH (adjusts the pH to ~7.0).
6. Autoclave for 20 min within 2 hr.
7. Store at room temperature.

Notes:

LB Agar Plates and Addition of Antibiotics

Make sure that the chemicals below, including 5 M NaOH, and the auto-clave are available before starting. Follow the autoclaving instructions from the lab teachers.

Introduction

These plates contain solidified lysogeny broth (LB), a rich growth medium for *E. coli*. In contrast to growth in liquid LB, where the bacteria are mobile, growth on plates gives colonies originating from one single bacterial cell.

Just before pouring the solution into Petri dishes, an antibiotic can be added for resistance selection. Normal working concentrations are:

- ampicillin: 100 µg/mL;
- chloramphenicol: 25 µg/mL;
- kanamycin: 50 µg/mL.

Normal stock concentrations:

- 1000-fold higher than above, respectively.

Note: Chloramphenicol stock is dissolved in ethanol.

Components

- NaCl;
- Bacto™ tryptone;
- yeast extract;
- agar;
- ddH$_2$O;
- 5 M NaOH;
- 1000x antibiotic of choice.

Procedure

Add the following to a 1 L bottle:

1. 600 mL LB prepared fresh as above (non-autoclaved).
2. 9 g agar.

3. Shake. It is unnecessary to dissolve all solids now because autoclaving will do this.
4. Autoclave for 20 min within 2 hr.
5. Let it cool to ~40–50°C (touchable, so the antibiotics will not be destroyed by the high temperature).
6. Add 600 μL of 1000x antibiotic of choice (if any) and gently swirl the bottle to mix. Do not shake the bottle vigorously as this will create many bubbles that will be transferred to your plates!
7. Pour into empty Petri dishes just enough to cover the surface (~20 mL per plate). If bubbles remain in the plates, heat the plate surface carefully with a burner to burst them. But make sure not to heat the solution in the plate too much since it might degrade the antibiotic.
8. Leave the plates at room temperature to solidify (~1 hr).
9. Solidified plates should be turned upside down for a few hours at room temperature, then stored at 4°C.

Notes:

Protocol 2. Overnight Cultures with Antibiotics and Glycerol Stocks

Overnight Cultures with Antibiotics

Setting up an overnight culture of a single bacterial strain needs a plate with single colonies and LB containing the appropriate antibiotic.

Introduction

Antibiotics should not be added to LB before autoclaving the LB or after subsequent cooling to room temperature for storage because antibiotics are not sufficiently stable. Antibiotic solutions are stored pure at −15°C and only added just before the addition of bacteria.

Components

Normal working concentrations are:

- ampicillin: 100 µg/mL;
- chloramphenicol: 25 µg/mL;
- kanamycin: 50 µg/mL.

Normal stock concentrations:

- 1000-fold higher than above, respectively.

Note: Chloramphenicol stock is dissolved in ethanol.

Procedure

1. Quickly burn the neck of a bottle containing LB medium before pouring it out into a flask (or tube). Even the slightest contamination of LB will be visible the next day as an unwelcome culture!
2. Add the antibiotic to give the appropriate concentration listed above.
3. Scoop one colony from the plate with a sterile inoculation loop (or micropipette tip).
4. Immediately stick the loop (or tip) into the flask (or tube) containing the medium and antibiotic for a few seconds.

5. Cover the flask loosely with aluminium foil (or cap tube loosely) by taping it on to allow exchange of oxygen. Good oxygenation is required, not only for maximal growth, but also for chemical maturation of the chromoprotein chromophores.
6. Incubate at 37°C with shaking overnight.

   Note: Check the temperature of your "37°C" room, as different or uneven temperatures can adversely affect results!

Cell Glycerol Stocks

Introduction

Cell glycerol stocks are stored at −80°C with the glycerol helping to maintain the viability of the cells.

Procedure

1. Mix 600 µL of an overnight culture with 400 µL of 50% glycerol (to give 20% glycerol final).
2. Place in the −80°C freezer.

Note: In contrast to the procedure for storing certain enzymes, a cell glycerol stock should not be cooled by snap freezing in liquid nitrogen.

Plasmid Preparations

Optionally, plasmids can be harvested from overnight cultures with a plasmid miniprep kit according to the manufacturer's instructions.

Notes:

Protocol 3. Biobrick™ 3A Assembly and Gel Analysis

Introduction

The starting point of every 3A assembly is plasmid DNA preps of three different BioBricks™ (top of Fig. 25). The endpoint is, hopefully, colored bacterial colonies.

Components

- DNA plasmids;
- ddH_2O;
- 10x reaction buffer for restriction enzymes provided by manufacturer;
- EcoRI restriction endonuclease, heat-inactivatable;
- XbaI restriction endonuclease, heat-inactivatable;
- SpeI restriction endonuclease, heat-inactivatable;
- PstI restriction endonuclease, heat-inactivatable;
- T4 DNA ligase;
- 10x reaction buffer for T4 DNA ligase provided by manufacturer;
- papers of every different color, as well as black, grey and white.

Procedure

Digestion

1. Make three mixes: each contains 500 ng of one of the three plasmids and ddH_2O to 43 µL. (Make four mixes when cloning into both high and low-medium copy vectors.)
2. To each mix, add 5 µL of 10x reaction buffer for restriction enzymes.
3. Add 1 µL each of the appropriate endonucleases (two per tube), according to Fig. 25, to give a final volume of 50 µL.
4. Tap on the tubes to mix. If necessary, centrifuge for a few seconds to spin down the liquid.
5. Incubate at 37°C for 30 min.
6. Withdraw a 20 µL aliquot for gel analysis (see below). Incubate the remainder at 80°C for 20 min if your brand of enzymes are heat-inactivatable.

At this point, samples may be stored at −20°C.

Gel analysis of digests (recommended for first time)

Run 20 μL of each digestion mixture (200 ng) on a 1% agarose gel (Protocol 4) to measure the extent of digestion. Also run the uncut plasmids (negative controls) directly beside their cut versions, and a DNA ladder marker should be loaded in a middle lane.

During the run, calculate the expected sizes of your products. Afterward, there should be one or more DNA bands visible under UV light in each lane. The marker will help indicate the size of the linear (cut) fragments. The lanes with cut plasmid should contain two bands: the slower one is the vector and the faster one is the insert. If the insert is very small (e.g. some promoter parts), it may be invisible due to a low amount of stainable DNA and low resolution, or due to running off the gel. The uncut plasmid should be one or two bands: a supercoiled version and a nicked form (one or both of the circular strands of DNA has a break in the phosphodiester chain, thereby allowing relaxation of the coil). Note that the different topologies of linears, nicked circles and supercoils cause each to migrate at a different rate. Beware of an unexpectedly slow migration of an uncut plasmid: this indicates a multimeric vector! It is also possible to have multiple different plasmids per cell.

Ligation

1. Add 2 μL (20 ng) of each of the three digestion mixtures to 11 μL of water.
2. Add 2 μL 10x reaction buffer for T4 DNA ligase.
3. Add 1 μL of T4 DNA ligase to give a final volume of 20 μL.
4. Incubate at room temperature (~22°C) for 30 min.
5. Heat-inactivate the enzymes by heating at 80°C for 20 min.

At this point, samples may be stored at −20°C.

Transformation

1. This first requires the preparation of competent cells (Protocol 5).
2. Competent cells are transformed as described (Protocol 6).

3. The next day, bacterial colonies are evaluated by eye for color. Try placing papers of every different color, as well as black, grey and white, underneath the agar plates to test which background color gives optimal sensitivity for detecting your faintly colored colonies.

Notes:

Protocol 4. Agarose Gel Electrophoresis

Be familiar with the safety procedures in Chapter 4 before starting. Preferably set up the gel equipment at multiple locations to avoid congestion.

Introduction

Agarose gel electrophoresis is used for separation and analysis of larger (>100 bases in length) nucleic acids under non-denaturing conditions. By adding the sample with loading buffer to the gel wells and applying a current over the anode and cathode, the negatively charged nucleic acid will migrate to the positive electrode. Analysis requires that the gel contains a DNA stain visible under UV light (see Chapter 4). Since the stain interacts with nucleic acids and is therefore potentially mutagenic, always wear nitrile gloves when working with agarose gels. If the stain is ethidium bromide, not Sybr®Safe, make sure that contaminated waste (pipette tips, gels, etc.) is disposed of in hazardous waste boxes. Use protective glasses when using the UV light box.

Components

- Agarose;
- 1x TBE (or TAE);
- Sybr®Safe;
- Loading dye mix;
- DNA ladder size marker;
- DNA samples.

Procedure

Casting a gel

1. Close the ends of the gel tray either with tape or by a casting stand, depending on the equipment. The gel tray must be on a level surface.
2. Insert the comb into the gel tray at one end ~1 cm from the edge.

3. For a 1% 150 mL agarose gel, weigh 1.5 g of agarose in a 500 mL conical flask. Add 150 mL 1x TBE (or TAE) buffer. If the gel tray is smaller, calculate the amount needed.

4. To dissolve the agarose in the buffer, swirl to mix and microwave for a few minutes, taking care not to boil the solution out of the flask. Remove the flask occasionally and check whether the agarose has dissolved completely. Be careful — the solution is very hot! Insulated gloves are too bulky to easily pull the flask out of the microwave, so a folded-up paper towel is suggested.

5. Let the agarose solution cool down. Once the solution is touchable, add the DNA stain. Check the stock concentration and add the appropriate amount to give the desired final concentration. The working concentration for ethidium bromide is 0.5 µg/mL, while for Sybr®Safe, it is simply 1x.

6. Pour the gel solution into the gel tray. Remove any air bubbles with a pipette tip. If not already in place, put in comb.

7. The gel will solidify while cooling down to room temperature. Depending on the initial temperature, this will take ~30 min.

Running the gel

1. Release the gel tray from the tape or casting stand. Place the gel tray into the buffer chamber and remove the comb carefully.

2. Add 1x TBE (or TAE) buffer until the gel is completely covered.

3. Take part of your DNA samples (~0.2 µg) and mix with loading dye. This can be done either in 1.5 mL tubes or, if the volumes are very small, on a piece of parafilm.

4. Load the size marker mixed in 1x loading dye (~6 µL final volume) into a middle well.

5. Load your samples into the other wells while writing down which lanes have which samples.

6. Put the lid onto the buffer chamber and connect it to the power supply. Make sure to put it in the right direction so that your DNA runs toward the positive (red) electrode.

7. Run the gel at 100 V for 30–60 min. Neither of the two dyes should be run off the gel. If the electrophoresis runs correctly, you will notice air bubbles coming from the negative (black) electrode.

8. Stop the run and bring your gel to a UV table to visualize your gel bands. Use protective glasses. If sufficient separation was not achieved, put the gel back into the buffer chamber and run it for longer.

9. Take a picture of your gel.

Notes:

Protocol 5. Preparation of Competent *E. coli* Cells Using CaCl$_2$

A plate with single colonies of a wild-type strain of *E. coli* will be provided at the start. In case this procedure fails, teachers should have a backup stock of competent cells available until students successfully make their own competent cells.

Introduction

E. coli is inefficient at taking up foreign plasmids. One cannot rely on its natural competence (ability to take up foreign DNA). For increasing the competence, the cell wall is made permeable by treatment with CaCl$_2$. It is important that the whole process is kept chilled! Remember to be careful with the cells as they become very fragile; e.g. resuspend cells in microfuge tubes by gentle micropipetting or tapping, not vortexing! *E. coli* DH5α is strongly recommended because of high competence and low ability to degrade foreign DNA by restriction at unmethylated sites.

Components

- SOB medium (LB is an acceptable substitute);
- plate with single-cell colonies;
- 0.1 M CaCl$_2$, ice-cold;
- 0.1 M CaCl$_2$ with 20% glycerol, ice-cold;
- liquid nitrogen.

Procedure

1. Take one colony and start a 5 mL overnight culture at 37°C, with shaking.
2. Dilute the overnight culture 1:100 into 50 mL SOB medium.
3. Grow culture at 37°C with shaking to an OD$_{600}$ = 0.4.
4. Let the culture sit on ice for ~15 min, swirling occasionally. When the cells are properly chilled, proceed to the next step.
5. Pour the culture into a 50 mL Falcon™ tube.

6. Centrifuge at 3500 rpm for 5 min at 4°C.
7. Remove as much as possible of the supernatant without disturbing the pellet.
8. Resuspend the pellet in 100 μL ice-cold 0.1 M $CaCl_2$ with the help of a sterilized loop.
9. Add 15 mL ice-cold 0.1 M $CaCl_2$. Mix gently by pipetting up and down a few times. Do not vortex!
10. Incubate the cells on ice for 30 min.
11. Pellet the cells again at 3500 rpm for 5 min at 4°C.
12. Resuspend the cells in 2 mL ice-cold 0.1 M $CaCl_2$/20% glycerol.
13. Incubate for 45 min on ice.
14. Aliquot carefully in 50 μL amounts to chilled 1.5 mL tubes.

Note: Some gentle mixing is required as cells tend to fall to the bottom of the liquid. Also, competent cells give the highest transformation efficiencies when fresh compared with after freezing in Step 15.

15. Snap freeze in liquid nitrogen any tubes that will not be used for transformation within a few hours.
16. Store at −80°C.

Note: Once the cells have been frozen and thawed, they cannot be refrozen again because this kills the cells.

Protocol 6. Transformation of CaCl$_2$-Competent *E. coli* Cells

This requires competent *E. coli* cells from Protocol 5 or supplied by the teacher.

Introduction

Remember that competent cells are very fragile and should only be mixed gently. Transformation plating results, like most results, can be very difficult to interpret without positive and negative controls. A good way to plan controls is to think how you might interpret different kinds of possible results. For example, are colonies really due to correct clones, or are they due to contaminating cells (due to degradation of the antibiotic plates stored for too long), contaminating plasmids or uncut vector? Is a lack of colonies due to failed ligations or just poor competent cells? **Three plating controls** are generally used to distinguish the above possibilities:

1. Water (a negative control).
2. Unligated, cut plasmid(s) (a negative control).
3. Intact plasmid (a positive control). This transforms much more efficiently than ligated plasmid, so 0.1 and 10 ng is sufficient. This enables calculation of the transformation efficiency of the competent cells. An additional advantage of this control is to compare the color intensities of the colonies. Thus, when cloning in the destination vector, use the destination vector as the control. When mutating a chromoprotein gene by PCR, use the plasmid containing the chromoprotein gene as the control.

Components

- CaCl$_2$-competent *E. coli* cells;
- SOB media (LB is an acceptable substitute);
- DNA plasmid.
- Agar plates.

Procedure

1. Turn on a water bath or heating block to 42°C.
2. Thaw competent cells on ice for 15 min.
3. Add 5 µL of ligation reaction mixture or controls above to 50 µL of competent cells.
4. Incubate for 30 min on ice.
5. Heat shock for 45 s at 42°C.
6. Incubate for 5 min on ice.
7. Add 950 µL of SOB media (pre-heated to 37°C).
8. Incubate for 1–1.5 hr at 37°C, with occasional gentle mixing by inversion of the tubes.
9. For positive controls, mix gently and plate 100 µL only (= 1/10th) on an agar plate containing the appropriate antibiotic as in Step 12.
10. For all samples, spin down cells from the original 1000 µL or remaining 900 µL at 4000 rpm for 5 min.
11. Discard all but 100 µL of the supernatant and resuspend the pellet in the remaining 100 µL.
12. Spread the remaining cell suspension on an agar plate containing the appropriate antibiotic as follows (see fire and burner safety procedures in Chapter 4 before you start):
    (i) Dip the spreader into 95% ethanol.
    (ii) Put it into the flame for a second.
    (iii) Let the ethanol burn off outside the flame.
    (iv) Spread the bacterial suspension evenly out on an agar plate. Continue until all the inoculum has gone into the agar.
    (v) Put the plates at 37°C overnight.

Following days

13. Day 2: Calculate the transformation efficiency (colonies/µg) of the competent cells using the positive control plates. How does it compare with the expected efficiency? In the evening, re-streak appropriate colonies (Protocol 7).
14. Day 3: Set overnight cultures.
15. Day 4: Make glycerol cell stocks (Protocol 2) of strains worth saving and process the rest of the culture according to instructions.

Notes:

Protocol 7. Bacterial Re-Streaking Techniques

It is worth practicing this method as early as possible in the lab because, although it only takes a few minutes, it takes almost a day to obtain the results and it can be tricky at first.

Introduction

Growth on plates gives colonies originating from one single bacteria cell. However, a single colony is not guaranteed to be clonal, and this is particularly problematic when there are several hundred colonies on a plate. Also, some colonies might not actually be resistant to the antibiotic in the plate (e.g. if they sit next to a colony secreting the ampicillin resistance enzyme). Re-streaking is recommended to address these issues, although it has the disadvantage of taking an additional day. There are many different re-streaking techniques, four of which are given here. Find out which procedure works best for you, or come up with your own method!

Procedure Alternative 1 (Figs. 27 and 28)

1. Take one colony from a plate using an inoculation loop and smear it out on one edge of a new plate.
2. Draw one line vertically through the first lines to the centre of the plate.
3. Sterilize the loop and then go back and forth through the vertical line all the way to the other edge of the plate.

Procedure Alternative 2 (Fig. 36)

1. Take one colony from a plate and streak it out within one quadrant at one edge of a new plate.
2. Sterilize the loop and then draw one line through the first set of lines.
3. Streak new lines at an adjacent quadrant of the plate.
4. Repeat the last procedure twice more to fill all four quadrants.

Procedure Alternative 3

This is similar to Procedure 2 but brings fewer cells to the next streak, more quickly reaching single cells on the plate.

1. Take one colony from a plate and streak it out within one quadrant at one edge of a new plate.
2. Sterilize the loop and start drawing the next line from the end of the previous streak.
3. Repeat the last procedure twice more to fill all four quadrants.

Procedure Alternative 4

This method eliminates many cells so that you can re-streak up to four different colonies on the same plate.

1. Divide a new plate into four labeled quadrants by marking the back of the plate.
2. Take one colony from a plate and swirl the loop around carefully in ~20 µL ddH$_2$O in a 1.5 mL tube.
3. Draw to and fro in one quadrant on the plate.

Notes:

Protocol 8. Lysis of *E. coli* Cells with Lysozyme and Freezing

The teacher may prepare the Tris-HCl and EDTA stocks beforehand because they require a pH meter.

Introduction

Lysozyme is an enzyme that digests the cell wall, with the main commercial source being hen egg white. Triton®X-100 is a detergent that solubilizes the cell membrane lipids. This protocol is adapted from Sambrook and Russel (2006).

Components

- 1 M Tris-HCl, pH 8.0;
- 0.5 M EDTA, pH 8.0;
- NaCl;
- Triton®X-100;
- lysozyme stock solution at 20 mg/mL in 10 mM Tris-HCl, pH 8.0;
- plate containing single colonies.

Procedure

1. Start overnight cultures of appropriate test and control strains of *E. coli*.
2. Prepare the lysis buffer from the first four components in the list above to give these final concentrations:
   10 mM Tris-HCl, pH 8.0;
   1 mM EDTA;
   100 mM NaCl;
   0.5 % (v/v) Triton®X-100.
3. Pellet 2 mL of an overnight culture in a 1.5 mL tube by centrifugation twice at 5000x g for 5 min. Remove the supernatants.
4. Resuspend the cell pellet in 300 μL of the lysis buffer.
5. Add 25 μL of lysozyme stock solution.

6. Mix by vortexing for a couple of seconds.
7. Incubate the sample at 37°C for 30 min.
8. Freeze-thaw 3 times to increase lysis.
9. Centrifuge the sample at maximum speed for 3 min to pellet the cell debris. Check the color of the pellet as it may tell the effectiveness of the cell lysis.
10. Take the supernatant and quantify its color using a spectrophotometer. If a 1 mL cuvette is used, this will require dilution with lysis buffer.

Notes:

Protocol 9. Polymerase Chain Reaction (PCR)

This protocol introduces PCR and explains how to order and dissolve primers. Protocols 10 and 11 below give details of specific PCR reaction mixes and programs in our workflow.

Introduction

PCR is an in vitro method for amplifying and/or mutating DNA (Fig. 31).

Components

- DNA template;
- 2 oligodeoxyribonucleotide primers;
- 4 dNTPs (dATP, dCTP, dGTP, dTTP);
- optimized buffer;
- thermostable DNA polymerase.

Choice of Thermostable DNA Polymerases

Fast polymerases, like the standard Taq polymerase, have a high error rate and do not extend well over long distances. In addition, Taq polymerase usually adds an untemplated A onto the 3' end of the PCR product. For amplifying long fragments, select a polymerase such as Phusion®HF with high processivity and also proofreading ability to increase fidelity.

Primer Design

The first step is to design and order primers that will work optimally. Primers may be completely complementary to their target template or have 5' overhangs, e.g. to introduce a restriction site, promoter or mutation (see Protocol 10). Poorly designed primers are the main cause of PCR reactions with incorrectly sized products or no products at all. Primers that hybridize at incorrect positions in your plasmid or in contaminating *E. coli* chromosomal DNA can be avoided by doing a **BLAST program** search to identify all positions in the target DNA that are complementary to a particular primer sequence, but this search is usually unnecessary. Other parameters to take into account are ensuring low probabilities of self-annealing (both intra- and inter-molecular) and annealing to the other primer (primer dimer-formation), which is checked using programs such as the **CLC Main Workbench program**. Although this program is expensive, a free trial version lasting one month can be downloaded by each student: http://www.clcbio.com/products/clc-main-workbench/. A similar arrangement exists for the **Snapgene program**. The hybridizing region should be at least 18 nucleotides, contain 40–60% G and C overall, and the 3′ ends should not have too high a GC-content. The calculated initial melting temperature (T_m) should be between 53°C and 62°C, and both primers should have similar melting temperatures. **This calculation should include information about the polymerase**: different polymerases have different buffers which have different salt concentrations which, in turn, affect T_m. In addition to the CLC Main Workbench program, there are several free programs available for calculating T_m (e.g. ThermoFisher T_m Calculator).

Dissolving Oligos

Ordered primers lack terminal phosphate groups and will be received dry in microfuge tubes. Paste the primer information into your lab notebook.

1. First, spin down the tube in a tabletop microcentrifuge at maximum speed for 3 min.
2. Look for the DNA pellet at the bottom of the tube. If it is invisible, assume that it is located there.
3. Add sufficient ddH$_2$O to the bottom of the tube to give a 100 μM stock.
4. Rinse the inside of the tube walls with the water and a pipette tip to ensure that all the DNA is dissolved. Dissolving is not instantaneous, so let the tube stand at room temperature for a few minutes until placing on ice.
5. Store primer stocks at −20°C.
   These stocks are ready to dilute for some PCR applications, but other applications such as inverse PCR mutagenesis (Protocol 10) require enzymatic phosphorylation of the 5′ ends.

PCR Reaction

First, find a PCR machine with the program desired or enter your own program based on the polymerase manufacturer's protocol, Protocols 10 or 11, and your primers. A typical PCR program is as follows:

1. Melting step, to separate the DNA template strands from each other.
2. Main cycle including:
   - melting step
   - annealing step
   - extension step
3. Repetitions of main cycle 20–30x.
4. Storage step.

Standard controls for PCR are:

1. Reaction without template (a negative control for contamination).
2. Positive control with 2 primers of Fig. 24A.

Notes:

Protocol 10. Inverse PCR Mutagenesis

Introduction

Inverse PCR mutagenesis (Hemsley *et al.*, 1989) is a flexible and rapid method for plasmid mutagenesis that combines PCR and cloning (Fig. 32). It enables the rapid alteration or substitution of a sequence up to 80 bases long, insertion of up to 80 bases, or deletion of almost any length desired. This protocol uses purified plasmid DNA target, but it is also possible to perform inverse PCR directly on a bacterial colony (see Protocol 11).

Primer Design

Follow the method used for PCR (Protocol 9) as closely as possible, mindful that compromises may be necessary due to the sequence of the target site. Primer length, including the 5′ overhang, should not be longer than ~60 bases to minimize mutations caused by the low fidelity of commercial chemical DNA synthesis. The key design features are that the two primers will extend away from each other and the two 5′ overhangs will become joined together at the mutation site in the gene (Fig. 33A).

Phosphorylation of Primers

Note: The protocol says ATP (ribonucleotide) and not dATP (deoxyribonucleotide).

Introduction

Commercially synthesized primers lack terminal phosphate groups. This is fine for some PCR applications, but for inverse PCR mutagenesis, a 5′ phosphate group is necessary for incorporation into a phosphodiester bond by T4 DNA ligase to circularize the PCR product into a plasmid.

Components

- ddH_2O;
- primer stock, 100 μM;
- 10x reaction buffer A for PNK provided by the manufacturer;

- ATP, 10 mM;
- T4 polynucleotide kinase (PNK).

Procedure

1. Prepare the following reaction mixture:

ddH_2O	70 μL
100 μM (100 pmol) primer	5 μL
10x reaction buffer A for T4 polynucleotide kinase	10 μL
10 mM ATP	10 μL
10 U/μL T4 polynucleotide kinase	5 μL
Total (5 μM primer ready to use)	100 μL

2. Mix thoroughly, spin down and incubate at 37°C for 30 min.
3. Heat-inactivate the enzyme at 80°C for 20 min.

Inverse PCR with Phusion®HF DNA Polymerase

Introduction

Phusion®HF DNA polymerase has a little endonuclease activity, so the incubation time with the DNA should not be extended. The program is a version of touchdown PCR that increases specificity but still provides a good yield. Two different PCR programs are recommended because long and short primers tend to have annealing temperatures above and below the extension temperature, respectively, in later cycles. The second program ensures that short primers can anneal for the extension reaction.

Components

- ddH_2O;
- dNTPs, 2 mM;
- phosphorylated forward primer, 5 μM;
- phosphorylated reverse primer, 5 μM;
- 5x Phusion®HF buffer provided by manufacturer;
- Phusion®HF DNA polymerase;
- DNA template.

Procedure

1. Choose one of the appropriate PCR programs (i) and (ii) below:
   (i) PCR program if you have **long overhangs**: the differences between this and the following program are highlighted in bold.

Description of Step	Temp. (°C)	Time (hh:mm:ss)	Number of Cycles
Initial denaturation	98°	00:05:00	1x
Denaturation	98°	00:00:30	
Annealing temp. + 4°C	°	00:00:30	2x
Extension	72°		
Denaturation	98°	00:00:30	
Annealing temp. + 2°C	°	00:00:30	2x
Extension	72°		
Denaturation	**98°**	**00:00:30**	
Annealing temp.	**°**	**00:00:30**	**6x**
Extension	**72°**		
Denaturation	**98°**	**00:00:30**	**25x**
Extension	**72°**		
Final extension	72°	00:07:00	1x
Storage	4°	∞	1x

   (ii) PCR program if you have **short overhangs** where the T_m of the combined initial base-pairing and overhang sequences does not exceed 72°C.

Description of Step	Temp. (°C)	Time (hh:mm:ss)	Number of Cycles
Initial denaturation	98°	00:05:00	1x
Denaturation	98°	00:00:30	
Annealing temp. + 4°C	°	00:00:30	2x
Extension	72°		
Denaturation	98°	00:00:30	
Annealing temp. + 2°C	°	00:00:30	2x
Extension	72°		
Denaturation	**98°**	**00:00:30**	
Annealing temp.	**°**	**00:00:30**	**30x**
Extension	**72°**		
Final extension	72°	00:07:00	1x
Storage	4°	∞	1x

2. Calculate the extension time in your lab notebook:
   - Length of amplicon: _______kb.
   - Phusion® pol extension time per kb: 15–30 s/kb.
   - Extension time (length × extension time per kb):
     _______hh:mm:ss.
3. Fill in the extension time in your chosen program grid above.
4. Use the manufacturer's T_m calculator and Phusion®information sheet to fill in the annealing temperatures in your chosen program grid above.

   Note: The recommended annealing temperature is above or equal to T_m for Phusion® pol, depending on primer length, but below T_m for Taq pol!
5. Program the PCR machine (unless an identical program already exists in the machine).
6. Dilute a small amount of your plasmid to 1 ng/µL.
7. Mix the seven components below in a PCR reaction tube (0.2 mL tube, which is smaller than a typical microcentrifuge tube).

ddH$_2$O	23.7 µL
2 mM dNTPs	5 µL
Forward primer (5 µM)	5 µL
Reverse primer (5 µM)	5 µL
5x Phusion® HF buffer	10 µL
Plasmid dilution (1 ng/µL)	1 µL
Phusion® HF DNA polymerase	0.3 µL
Total	50 µL

8. Run the PCR reactions including controls (p. 107).
9. Analyse a 5 µL aliquot by agarose gel electrophoresis.
10. If a full-length band was visible, purify your PCR product using a PCR purification kit (or gel extraction kit) according to the manufacturer's protocol. However, elute the DNA in just 30 µL to maximize your chance of quantifying it (6 ng/µL sensitivity limit for Nanodrop™).

> ### *PCR purification kit*
>
> This purifies DNA products from PCR reaction components to enable downstream applications. Always consult the manufacturer's protocol because the procedure can differ between kits. Typically:
>
> 1. A high-salt solution is added to the PCR reaction mix.
> 2. The PCR solution is loaded onto a silica minicolumn.
> 3. The minicolumn is microcentrifuged to force the solution through.
> 4. A wash solution is loaded.
> 5. The minicolumn is re-microcentrifuged to remove contaminants.
> 6. A low-salt elution buffer is loaded.
> 7. The minicolumn is again re-microcentrifuged to elute the purified DNA.

11. Measure the concentration of your PCR product.

Digestion with DpnI

Note: Extended digestion times with DpnI may be beneficial.

Introduction

DpnI only digests methylated DNA sites (Fig. 33B), i.e. in vivo synthesized DNA like the original vector that was extracted from *E. coli* in a previous step. The PCR product will, on the other hand, be unmethylated and thus remain intact through the reaction.

Components

- ddH_2O;
- purified PCR product;
- 10x FastDigest buffer for restriction enzymes provided by manufacturer;
- FastDigest DpnI.

Procedure

1. Calculate the volume of the purified PCR product containing 500 ng and fill in the table below.
2. Calculate the volume of water to make up to 40 μL.
3. Mix the components below.

ddH$_2$O	μL
DNA from inverse PCR (500 ng)	μL
10x FastDigest buffer	1 μL
FastDigest DpnI	1 μL
Total	40 μL

4. Incubate at 37°C for 3 hr to overnight (despite the recommendation of only 15 min by ThermoScientific™).
5. Heat-inactivate the enzyme at 80°C for 20 min.
6. Use a PCR purification kit to purify the PCR product according to the manufacturer's protocol. However, elute the DNA in just 30 μL to maximize your chance of quantifying it (6 ng/μL sensitivity limit for Nanodrop™).
7. Measure the concentration of your PCR product.

Ligation

Introduction

The blunt ends on the inverse PCR product are ligated together by T4 DNA ligase to form a circular plasmid for transformation. Save some unligated, DpnI-treated, purified PCR product for transformation (a negative control).

Components

- ddH$_2$O;
- purified, DpnI-treated PCR product;
- 10x reaction buffer for T4 DNA ligase provided by manufacturer;
- T4 DNA ligase.
- 30% (w/v) PEG — adding to give 5% final concentration can help.

Procedure

1. Calculate the volume of the DpnI-treated PCR product containing 50 ng and fill in the table below.
2. Calculate the volume of water to make up to 50 μL.
3. Mix the components below:

ddH$_2$O	μL
Purified PCR product (50 ng)	μL
10x reaction buffer	5 μL
T4 DNA ligase, 5 U/μL	1 μL
Total	50 μL

4. Incubate for 60 min at room temperature (~22°C).
5. Heat-inactivate the enzyme at 80°C for 20 min.
6. Transform competent cells according to Protocol 6.
7. On the following day, re-streak appropriate colonies.

Notes:

Protocol 11. Colony PCR

Note that in contrast to Protocol 10, it does not matter whether or not the primers have been 5' phosphorylated. Also, Taq DNA polymerase is preferred for colony PCR.

Introduction

The primary goal of colony PCR (using primer sites shown in Fig. 24) is to prepare DNA samples from several colonies to screen for the desired mutant by sequencing. DNA samples for sequencing are more easily prepared by colony PCR reactions than by overnight cultures followed by plasmid preps. There is no need for a cell lysis step because the DNA template is released from the bacterial colony during a PCR reaction. Before sending for sequencing, PCR products are sized and quantified on an agarose gel to verify that the PCRs worked. Also, if a large deletion or insertion was desired, this size difference may be visible on the gel as a preliminary screening step before sequencing (provided that you also run a control PCR reaction for comparison!).

Components

- A plate with re-streaked, single colonies;
- ddH_2O;
- dNTPs, 2 mM;
- VF2 primer;
- VR primer;
- 10x Taq PCR buffer provided by manufacturer;
- Taq DNA polymerase.

Procedure

Screening requires testing several different colonies at once, i.e. several PCRs. In such cases, instead of preparing several reaction mixes individually, the number of pipetting steps can be reduced and the reproducibility of the reactions increased as follows: prepare just one "**master mix**" that includes all components except the variables (in this case, the cells). For example, if you decide to perform six PCRs (including your original control colony and a negative control), then you will need to make up a volume of master mix equivalent to seven reaction mix volumes, not six volumes, because there never seems to be enough volume in the last aliquot! So, instead of using the volumes given for the reaction mix in Step 6 below, you would use 7x as much of each volume to make up your master (pre) mix, then aliquot 49 µL six times into different 0.2 mL tubes. The remaining volume-deficient "seventh aliquot" is then discarded.

1. Calculate the extension time in your lab notebook:
   * Length of amplicon: _______ kb.
   * Taq extension time per kb: 1 min/kb.
   * Extension time (length × extension time per kb):
     _______ hh:mm:ss.
2. Fill in the extension time in the program grid:

Temp. (°C)	Time (hh:mm:ss)	Number of cycles
94°	00:05:00	1x
94°	00:00:30	
58°	00:00:30	30x
72°		
72°	00:07:00	1x
4°	∞	1x

3. Program the PCR machine (unless an identical program already exists in the machine).
4. Number the colonies you wish to test by marking on the backs of test and control plates.
5. Using a sterile loop, pick a small portion of each of these colonies and suspend individually in 30 µL of water. Mix thoroughly.

6. Calculate the volumes for making up your pre-mix for several reactions (as described above) based on the volumes provided for just one reaction here:

ddH$_2$O	28.7 μL
10x Taq PCR buffer	5 μL
2 mM dNTPs	5 μL
VF2 (5 μM)	5 μL
VR (5 μM)	5 μL
Taq DNA polymerase	0.3 μL
Total reaction mix	49 μL

7. Prepare your pre-mix and divide into 49 μL aliquots on ice.
8. Add 1 μL of each cell suspension to each aliquot and mix.
9. Run the PCR reactions.
10. Analyze a 5 μL aliquot of each PCR by agarose gel electrophoresis.
11. For reactions where a full-length band was visible, purify the PCR product using a PCR purification kit according to the manufacturer's protocol. However, elute the DNA in just 30 μL to maximize your chance of quantifying it.
12. Measure the concentrations of your PCR products.
13. Mix with primer VF2 for sequencing according to the company's instructions. Eurofins Genomics takes:
    15 μL PCR DNA (10 ng/μL) + 2 μL primer (10 μM).

Notes:

Protocol 12. Gibson Assembly

Introduction

This PCR-based method (Gibson *et al.*, 2009) joins multiple DNA parts in a manner independent of DNA sequence. The 5' ends of the primers are designed to include 20–150 base overlaps with the fragment that they will be ligated to (Fig. 38). The difficulty increases with the number of parts in the reaction; up to 12 were reported by Gibson *et al.* Each DNA can be up to several hundred kilobase pairs in length and should not be less than ~250 bp because of the risk of complete degradation by the exonuclease. Shorter fragments can be tried with less endonuclease.

Components

5x Isothermal Reaction Mix:

ddH$_2$O	mL
1M Tris-HCl, pH 7.5	3 mL
2 M MgCl$_2$	150 μL
1 M DTT	300 μL
100 mM NAD	300 μL
100 mM dATP	60 μL
100 mM dCTP	60 μL
100 mM dGTP	60 μL
100 mM dTTP	60 μL
PEG-8000	1.5 g
Total	6 mL

Aliquot in 320 μL portions and store at −20°C.

Assembly Master Mix:

ddH$_2$O	mL
5x Isothermal Reaction Mix	320 μL
10 U/μL T5 exonuclease	0.64 μL
2 U/μL Phusion® HF DNA polymerase (normal version, not hot-start version)	20 μL
40 U/μL Taq DNA ligase	160 μL
Total	1.2 mL

Aliquot in 15 μL portions and store at −20°C for a year or more.

Procedure

1. Perform PCRs (Protocol 9 with unphosphorylated primers) with your target DNAs and Phusion® HF DNA polymerase.
2. Gel purify the PCR products with a kit (Table 4).
3. Thaw a 15 μL aliquot of Assembly Master Mix.
4. Add 5 μL of ~equimolar mixture of three DNAs. Use 10–100 ng of each ~6 kbp DNA fragment.
5. Incubate at 50°C for 15–60 min, where 60 min is optimal.
6. Transform into competent cells (Protocol 6).

Trouble-shooting by Overlap Extension PCR

Some overlaps seem to be more problematic than others, perhaps due to secondary structure formation at 50°C (e.g. overlaps that contain terminator stem loops). If a particular junction is hard to obtain by Gibson assembly, it can first be obtained by **overlap extension PCR** (OE-PCR), which is PCR of the two or more fragments together without primers (Fig. 37). For OE-PCR, use enough DNA (>200 ng) to see bands on a gel to make sure it worked. Then use the joined fragment in a standard Gibson assembly.

Notes:

7

Advanced Methods

Flow Cytometry and Cell Sorting

Flow cytometry is a powerful technique for analyzing fluorescent microscopic particles such as bacteria and eukaryotic cells (Figs. 40 and 41).

Unlike a fluorometer that measures the average fluorescence of a population of cells, flow cytometry gives the actual distribution of fluorescence among individual cells in a population. The cells are suspended in a narrow stream of fluid such that only one cell passes through the laser beam at a time. Each time that happens, the cell scatters the light slightly, both in parallel with the beam (forward scatter) and perpendicular to it (side scatter). Also, if the cells contain fluorescent molecules, they will be excited into emitting light of a different color than the laser. A set of detectors in the instrument measures both light scattering and fluorescence, giving detailed quantitative information about individual cells in the population analyzed. Since thousands of cells can be analyzed per second, large amounts of data can be acquired from each sample, giving good

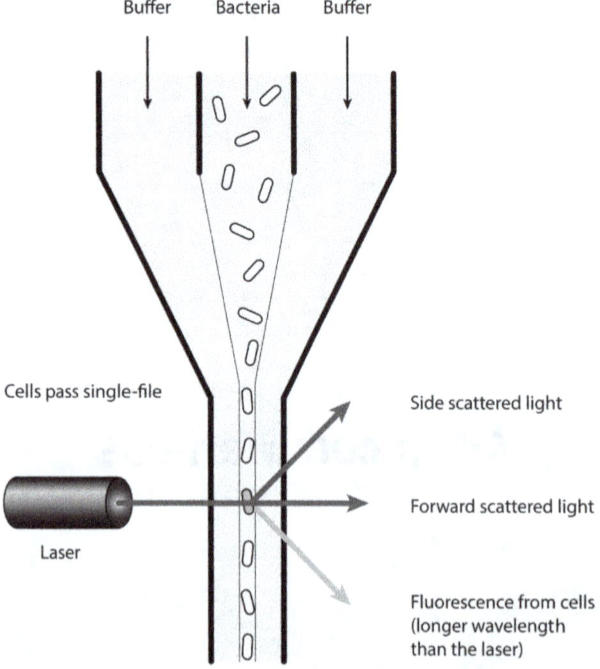

Fig. 40 Schematic of analysis of fluorescent cells by flow cytometry. With respect to chromoproteins, the technique only works with those that can fluoresce to some extent.

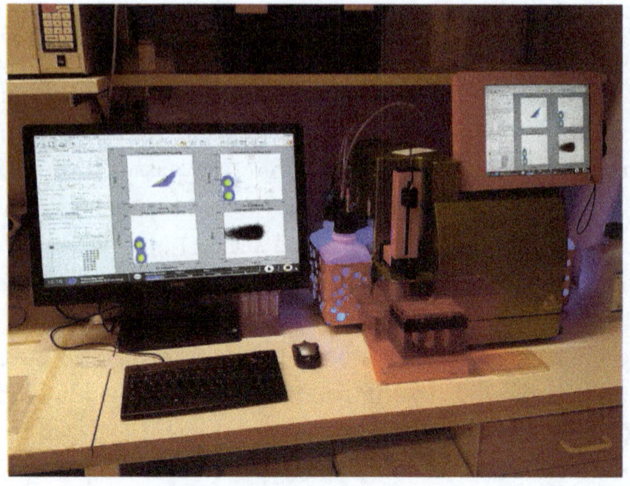

Fig. 41 Flow cytometer machine. It functions as illustrated in Fig. 40.

statistical accuracy of the measurements. Reporters such as GFP can be used both for labeling of cells and as reporters of gene expression. Fluorescent markers with different colors make it possible to measure the expression levels of several different genes in a single cell simultaneously or to track different types of cells in a mixed population. With respect to chromoproteins, the technique only works with those that can fluoresce to some extent: the only two in our chromoprotein parts being amilGFP and amajLime (Table 1 in Appendices). However, we also list two fluorescent proteins (Table 1 in Appendices), Super Yellow Fluorescent Protein 2 (SYFP2; Kremers *et al.*, 2006) and Blue Fluorescent Protein (mTagBFP; Subach *et al.*, 2008), as these are ideal for flow cytometry.

Some flow cytometers can sort the analyzed cells into different containers based on the optical characteristics of each cell, a technique called fluorescence-activated cell sorting (FACS; Fig. 42).

An application of FACS used by Uppsala iGEM students is diagrammed in Fig. 50.

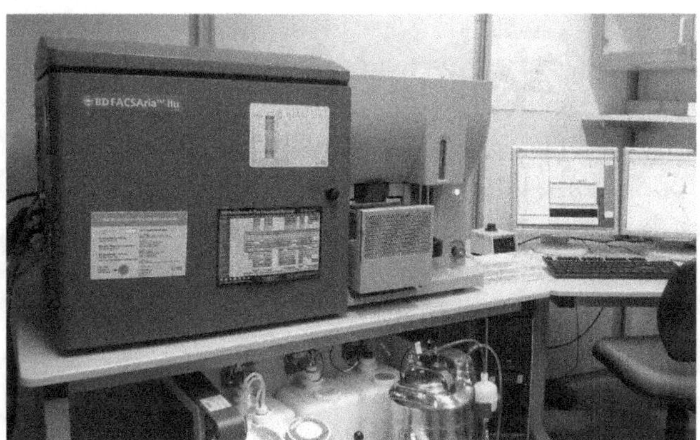

Fig. 42   Fluorescence-activated cell sorter (FACS) machine. This advanced flow cytometer both analyzes cells and sorts them into different containers based on their optical characteristics.

Recombination in Plasmids and the Chromosome

Recombination-based in vivo methods are essential for adding, subtracting or swapping genes on chromosomes, but they are also competitive with restriction-enzyme-based methods for manipulating plasmid DNAs.

Choosing between gene expression from plasmids versus the chromosome

There are many benefits of using plasmids when constructing genetic devices in bacterial chassis. Plasmids can be easily prepared from bacteria, they can be cut with restriction enzymes for screening and for assembly of different parts, they are easy to move between different bacterial strains, and plasmids with high copy numbers allow very high expression levels of genes when needed. Plasmids are also very useful for the storage and distribution of genetic parts. However, bacteria can sometimes lose plasmids, especially when they contain highly expressed parts or other constructs that are costly or even toxic for the cells. Plasmids can be more or less stable depending on their origin of replication but, in general, it is necessary to use constant antibiotic selection to ensure that the plasmids are not lost. Some plasmids can also have high variation in their copy number, which could be problematic if specific expression levels are critical.

An alternative to expressing genes from plasmids is to integrate them on the bacterial chromosome. This gives single-copy expression levels and higher genetic stability, obviating the need for antibiotic selection. Some integration methods require antibiotic resistance markers during the construction stage, but these can be removed later through the use of site-specific DNA recombinases (see below). Removal of antibiotic resistance markers is a requirement for some medical and environmental applications.

Lambda Red Recombineering

The word "recombineering" is derived from "recombination-mediated genetic engineering." Recombineering is a technique that can be used to "knock out" (Fig. 43), replace (Fig. 44), modify or insert genetic material on the bacterial chromosome or on a plasmid.

Unlike cloning and assembly methods that use restriction endonucleases and DNA ligases, recombineering does not require conveniently

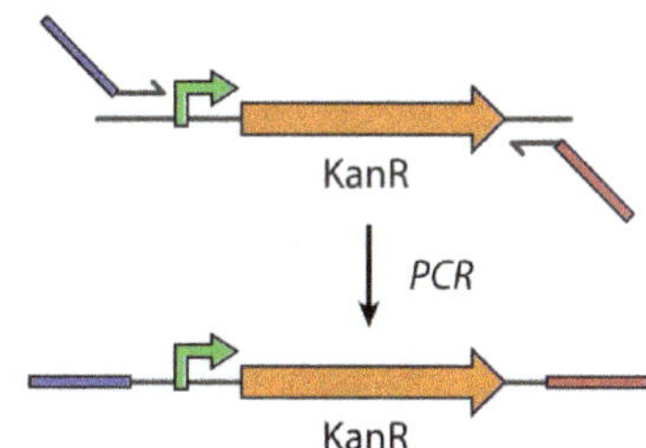

2. Electroporate into strain expressing Lambda Red system.

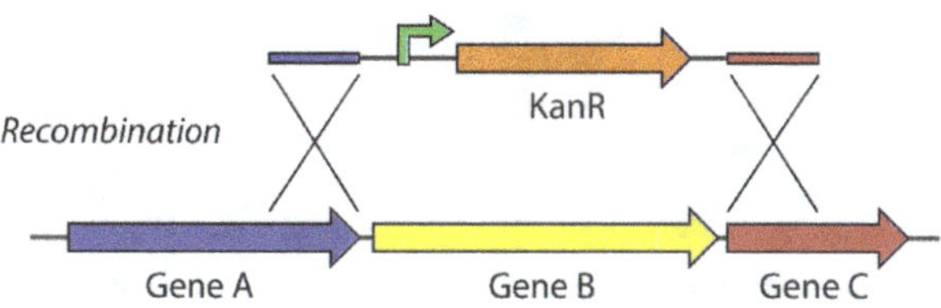

3. Positive selection of antibiotic resistant transformants.

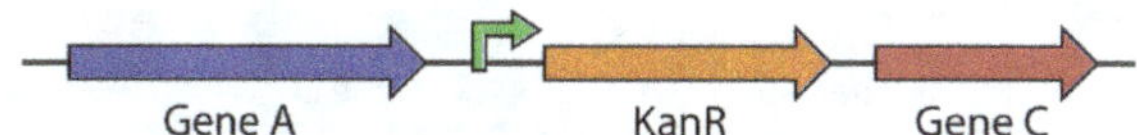

Fig. 43   Knocking out a gene (yellow) by recombination with an antibiotic resistance gene (orange).

positioned restriction sites; it is instead based on sequence homologies. This makes the technique similar to other homology-based methods such as Gibson assembly, but while Gibson assembly uses enzymes to join different sequences in vitro, recombineering is done in vivo. In general, the constructs to be integrated are generated by PCR using primers with overhangs containing homologies to the target location.

There are some limitations to recombineering systems, with perhaps the biggest one being the construct size. While the integration of 2–4 kbp is highly efficient, constructs above this size will have reduced integration frequency. To be able to select successful recombinants, the linear construct must contain a selectable marker, preferably a marker flanked by FRT or lox sites (see below) if the marker should be removed using site-

1. PCR amplify gene of interest together with resistance gene using primers with 40 bp homologies to the target.

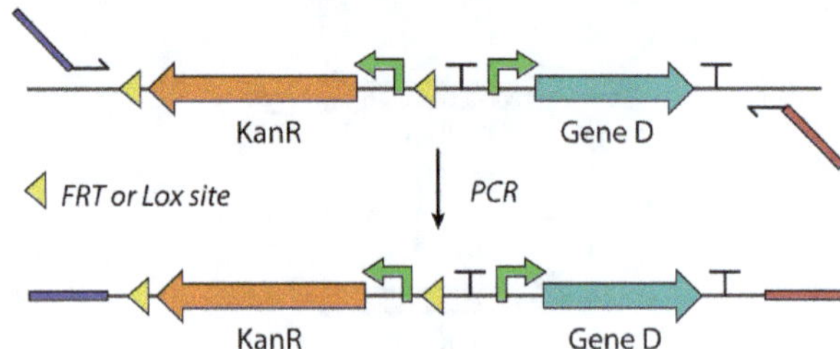

2. Electroporate the targeting construct into strain expressing the Lambda Red recombination system.

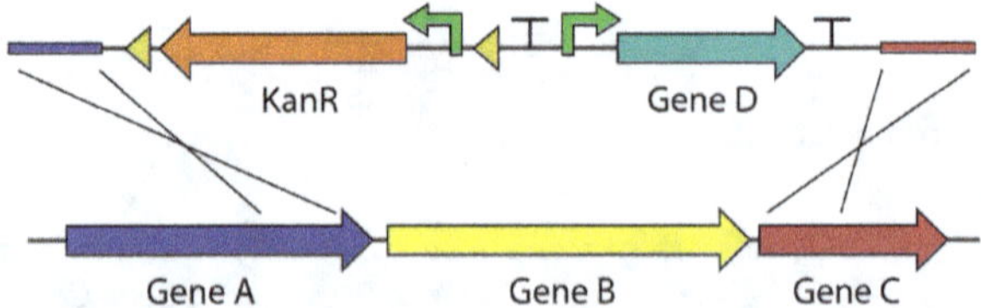

3. Positive selection of antibiotic resistant transformants.

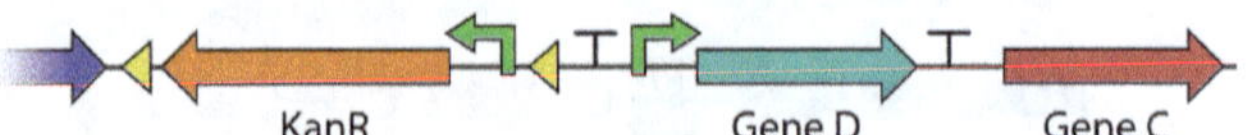

4. Eliminate resistance cassette using site specific recombination (FLP or Cre recombinase expression plasmid).

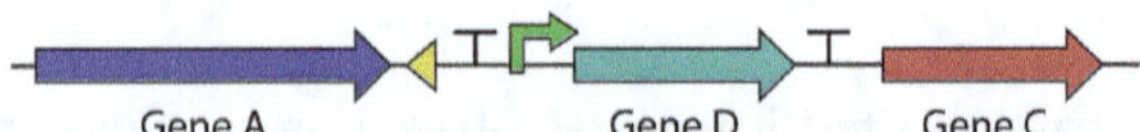

Fig. 44   Replacing a gene (yellow) by recombination with an antibiotic resistance gene (orange) and replacement gene (green). The resistance gene is then removed by site-specific recombination between the short sequences represented by two yellow triangles (see also Fig. 47).

specific DNA recombinases after integration of the construct. Since the integration method is based on homology, it is also critical that the ends of the linear construct are the only sequences homologous to the chromosome or any other genetic material present in the recipient cells. All homologies longer than 20–30 bp besides the targeting sites could give unpredictable integration, and this could be problematic since many commonly used standard parts such as terminators and promoters are derived from native *E. coli* sequences. When constructing systems that are intended to be integrated on the chromosome with recombineering, the use of synthetic parts or parts derived from other organisms could help avoid this problem.

How Lambda Red recombineering works

The Lambda Red recombineering system works in *E. coli* and *Salmonella* and is based on the three "Red" proteins from bacteriophage Lambda that promote homologous recombination:

1. Protein "Gamma" prevents the *E. coli* nucleases RecBCD from degrading the linear DNA.
2. Protein "Beta" binds to single-stranded DNA to promote annealing.
3. Protein "Exo" is a 5′ to 3′ exonuclease that chews back on the ends of the linear DNA to make it single stranded.

The genes for these three proteins can be expressed from a thermosensitive, low copy plasmid such as the pSIM-plasmids (Datta *et al.*, 2006). In these plasmids, the expression of the three Red genes is controlled by a temperature-sensitive repressor that allows tight repression at 30°C. Strains carrying these plasmids are grown at 30°C except when you want to integrate a new construct. Induction of the system at 42°C for 15 min is enough to make the bacteria express a pulse of Lambda Red proteins, making them able to integrate linear DNA. The cells are then made electrocompetent (see below), and the linear construct can be transformed into the bacteria by electroporation. Next, cells are allowed to recover for a few hours before plating on selective media, where only cells that have successfully integrated the synthetic construct will be able to survive. The pSIM plasmid carries a thermosensitive origin that can replicate only below 37°C, so to cure the strain from the plasmid after successful integration, the strain is grown at 42°C overnight. Another common method for chromosomal integration is the Tn7 transposon system.

Single-Stranded Recombineering and Multiplex Automated Genome Engineering (MAGE)

Although the Lambda Red method above uses double-stranded DNA generated through PCR, recombineering can also be done using synthetic single-stranded oligodeoxyribonucleotides. Unlike double-stranded recombineering, no exonuclease activity is needed to make the mutagenic DNA single stranded, and expression of just protein "Beta" is sufficient for efficient recombination (Fig. 45).

The mutation efficiency depends on both the type of mutation attempted and the oligo design. Since the mutagenic oligo will anneal to the lagging strand of the replication fork, it is important to design it to target the correct strand of the genome. Given that replication is bidirectional from the origin of replication, the strand targeted depends on the

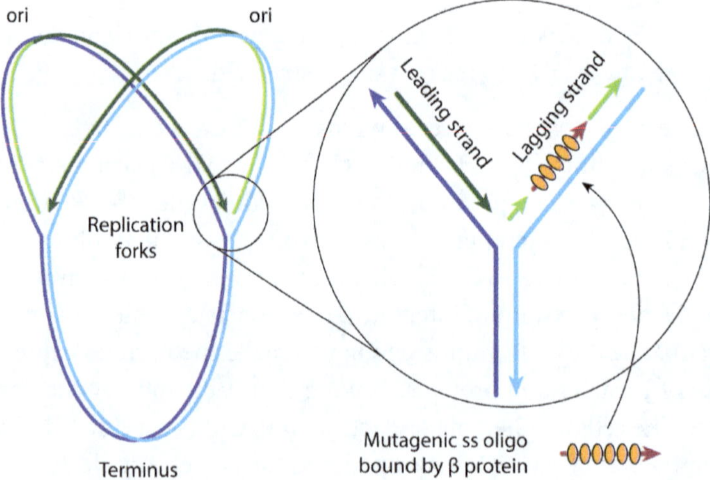

Fig. 45   Single-stranded recombineering. Left: chromosomal replication in *E. coli* is bidirectional, starting at the origin of replication (oriC). For single-stranded recombineering, it is important to target the strand templating synthesis of lagging-strand DNA. Right: Beta protein (orange) binds to an oligo (red) electroporated into the cell and promotes annealing of the oligo to a lagging-strand target location exposed on the chromosomal replication fork. The mutagenic oligo is then incorporated between Okazaki fragments into the nascent DNA.

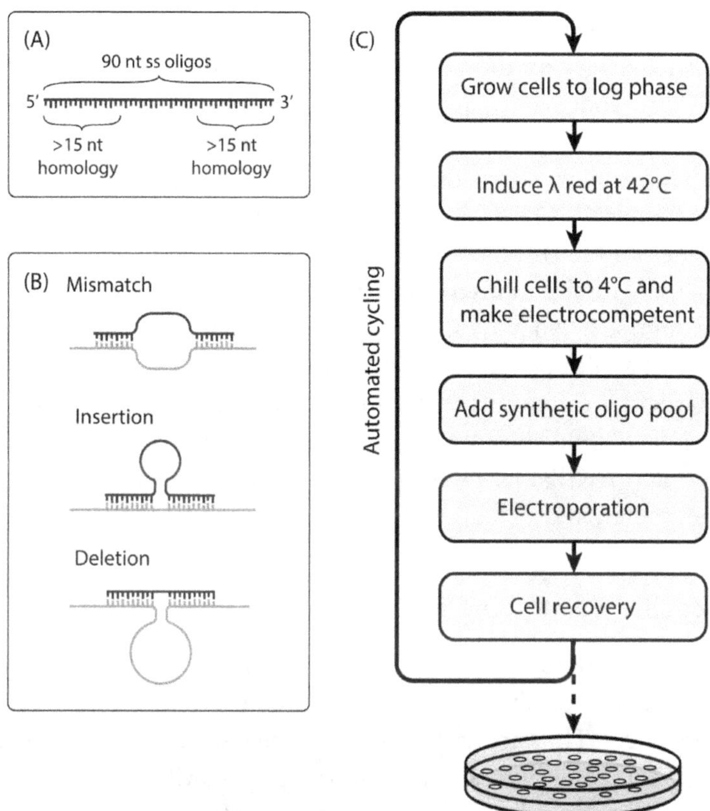

Fig. 46 Synthetic oligo design for single-stranded recombineering and multiplex automated genome engineering (MAGE). (**A**) Optimal oligo length and homologies with the target sequence. (**B**) Oligo designs for three types of mutations. (**C**) MAGE. The fraction of mutant bacteria in the population increases with each cycle.

position of the locus compared to the origin (Fig. 45). An efficient oligo is generally about 90 nucleotides long with at least 15-nucleotide homologies to the target at each end of the oligo (Fig. 46A).

The longer the homology to the target, the more efficient the incorporation will be. Nucleotides in the middle of the oligo introduce mutations by mismatch or insertion or deletion (Fig. 46B). Deletion efficiencies decrease with increasing deletion size. It is also important to consider secondary structures of the oligo, as strong hairpins lower efficiencies.

Efficiencies can be further increased by substituting the first four phosphodiester linkages on the 5' end of the oligo with phosphorothioates to protect from exonuclease degradation.

Single-stranded recombineering is much more efficient than the double-stranded version, so the single-stranded version has the advantage that it is independent of selectable markers and does not leave any "scars" such as FRT sites on the chromosome. However, compared to double-stranded Lambda Red, the size of insertion is limited by the size of the oligo.

The high efficiency of single-stranded recombineering has enabled fast, highly parallel, genomic engineering using a mix of many mutagenic oligos targeting different chromosomal loci and repeating many times (Wang *et al.*, 2009). This method is termed multiplex automated genome engineering (MAGE; Fig. 46C). Users should be aware of the likelihood of off-target mutations.

Site-Specific Recombination

Site-specific recombination techniques use a class of enzymes called recombinases that catalyzes the recombination between two short, specific DNA sequences. They are useful for removing a selection marker used for chromosomal integration and for constructing a switch that changes the direction of a promoter under controlled conditions in vivo. Two of the most common systems are the Cre/loxP system derived from a bacteriophage and the Flp/FRT system derived from yeast. The Cre recombinase will recognize sequences called loxP, while the Flp recombinase will bind to sequences called FRT ("flippase recognition target"). Recombinases are fast, specific and efficient, and genetic segments flanked by such sequences will either be excised (and lost) or inverted, depending on the orientation of the recognition elements (Fig. 47). Scarless methods have also been developed.

Other Recombination Methods

Three other common methods based on recombination should also be mentioned. Bacteriophage transduction is useful for moving large chromosomal fragments between bacteria. Yeast can be transformed with

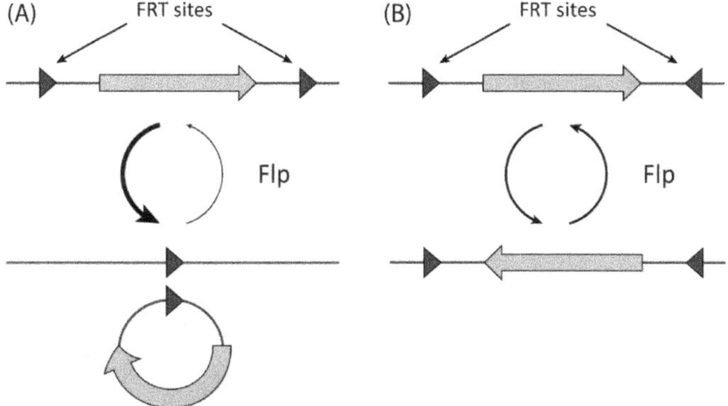

Fig. 47   Site-specific recombination by FRT recombinase. Depending on the orientation of the two short recognition elements (triangles), genetic segments flanked by them are either (**A**) excised (and lost) or (**B**) inverted.

small or large fragments for extra-chromosomal recombination (Oldenburg *et al.*, 1997) as an in vivo alternative to Gibson assembly. And CRISPR editing often involves recombination in vivo (see below).

Electrocompetent Cells

When using linear DNA fragments to transform *E. coli* during recombineering, higher transformation efficiencies are required than obtainable by the heat shock method, so electroporation is used instead. The protocol for making *E. coli* electrocompetent is very similar to the protocol for making them chemically competent, but the cells are washed with ice-cold distilled water or 10% glycerol instead of calcium chloride. The electrocompetent cells are then mixed with the DNA in an electroporation cuvette, and an electric pulse is passed through the cuvette using a voltage of ~2 kV (Fig. 48).

The cells are then allowed to recover and plated on selective agar, similar to transformation by heat shock.

CRISPR Editing

Genetic editing of eukaryotic genomes has been dramatically simplified recently thanks to novel synthetic biology tools based on bacterial defense

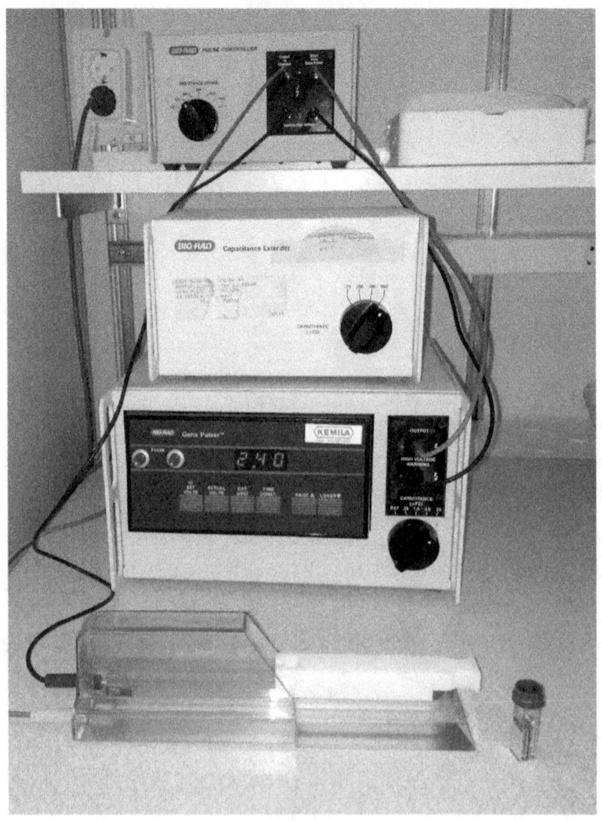

Fig. 48   Electroporation apparatus. The disposable/reusable cuvette on the right is pushed into the left-hand side of the container for administration of a high voltage.

systems termed CRISPR (clustered regularly interspaced short palindromic repeats).

Until a decade ago, genetic engineering of eukaryotic genomes was limited to model organisms and projects could take years (e.g. construction of knockout mice). The revolution came from complete characterization, simplification and generalization of a CRISPR system, such that a single RNA (crRNA-tracrRNA chimera) could be engineered to target virtually any dsDNA site for double-stranded cleavage by a natural protein called Cas9 (Jinek *et al.*, 2012; Fig. 49). The bonus is that the bacterial-

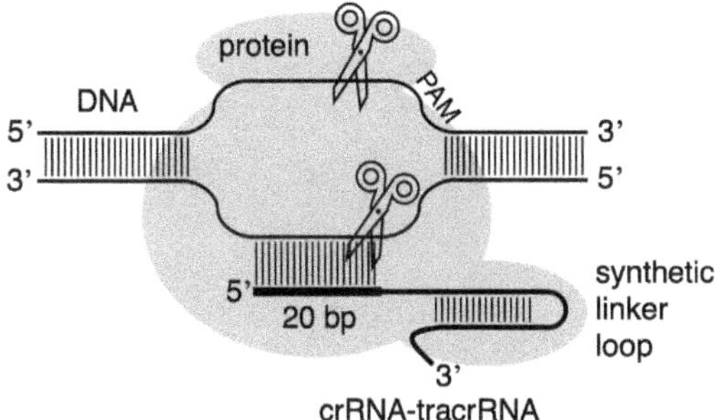

Fig. 49   Targeting DNA for double-stranded cleavage by Cas9 protein (grey) and a synthetic crRNA-tracrRNA chimera. Specificity is governed by perfect complementarity to a custom-designed 20-nucleotide sequence and also a 2–3 nucleotide sequence termed PAM (protospacer adjacent motif).

derived system works efficiently in vivo in a wide range of heterologous organisms, including plants and animals (Doudna and Mali, 2016). Although targeting requires a short (two base pair) adjacent PAM (protospacer adjacent motif) sequence and cleavage is imprecise, cellular repair mechanisms rescue the cleaved chromosome. This occurs by non-homologous end joining (disrupting the gene) or, with donor DNA, by a lower-efficiency homology-directed repair (correcting or inserting a gene).

This SynBio technology is advancing so rapidly that consultation of current literature is advised for the latest methods. CRISPR has enabled gene drives (conversion of both alleles in offspring), and CRISPR base editors and prime editors have been invented by David Liu's lab to directly make desired base changes, insertions or deletions. The system has also been coupled to transposases and epigenome editors, and delivery by lipid nanoparticles is progressing.

Unrelated methods using **"bridge"** RNAs to target recombinases for genome editing also look promising (Tou and Kleinstiver, 2024).

8

The international Genetically Engineered Machine (iGEM) Competition

The amazing phenomenon known as iGEM was introduced in Chapter 1, and three prize-winning early iGEM projects were summarized in Lab section 4 of Chapter 5. Here, we provide more practical information, including chronological blow-by-blow descriptions of the 2011 and 2012 Uppsala iGEM projects that inspired this manual. This illustrates how iGEM teams work from start to finish, provides the origins of the chromoprotein and antisense genes used in this lab course, and documents the follow up all the way to publication in scientific journals.

How to Start an iGEM Team

Detailed practical information on getting started is available from the iGEM webpages, such as http://igem.org/Start_A_Team and http://igem.org/Videos/Lecture_Videos.

Prior iGEM team members should be consulted as early as possible. They will not only show you the ropes but also likely advise that their team ran out of time because they did not start early enough or they did not expect commercial DNA synthesis to take so many weeks or their experiments did not work as planned.

Faculty should also be approached early too. iGEM fundraising and insurance problems were solved at Uppsala University by faculty offering to turn iGEM into an official university course, an atypical move for the iGEM competition. Although iGEM is, by its very nature, student-driven, this does not mean that instructors and advisors should not be exploited, especially ones in your chosen research area. Bounce your project brainstorming ideas off them, run your primer designs by them, and solicit their feedback on drafts of your wikis, posters and presentations early enough to make a real difference.

There is no single recipe for success. Students may pick a project largely because it is related to work performed in a research lab at the same institution or related to the institution's previous iGEM project. Such a project is less difficult to get (re-)started, get supported, get completed in the course of one summer, and even published. Publication is an important quality goal of iGEM, as only publication in a peer-reviewed scientific journal provides strong evidence for completion, novelty and utility, together with the availability of methods and materials sufficient for others to repeat and build upon it.

But if the project is closely related to existing research, student creativity will be less evident and, in reality, such a project may not be any easier to bring to fruition.

Uppsala iGEM 2011 — Show Color with Color

http://2011.igem.org/Team:Uppsala-Sweden

In spring 2011, the third straight Uppsala University iGEM team formed from 18 undergraduate students, students mostly enrolled in a five-year bachelor/master degree program in Molecular Biotechnology. For the first time, the competition was made into an official course in synthetic biology to guarantee work and travel insurance, partial university funding, and summer accommodation assistance for the students.

After many discussions amongst the students, the team decided to go for a project based on bacterial gene regulation by light. Compared to more traditional ways of inducing gene expression in bacteria, such as adding chemicals like IPTG, light is cheaper and can be regulated more precisely in time and space. To show a proof of principle, **the team planned to construct a bacterial photographic color film**. Although Uppsala was not working in this area, the basic concept had already been shown elsewhere with the first black-and-white "coliroid" photo (see "Picture this" in Lab section 4 of Chapter 5). The idea was later expanded to show that multichromatic control of gene expression was possible by using two **light sensors** reacting to two different wavelengths of light, **red** and **green**. The Uppsala team wanted to further expand this to enable complete spectral color readout by adding a third light sensor, just as our eyes and color TVs use three different colors of light. To make the bacteria behave like a photographic film, each light sensor, reacting to different colors of light, would be connected to a different reporter that would make the bacteria change color when illuminated.

The genes for two light sensors were obtained from the Voigt lab: the cph8 gene product reacts to red light and ccaS senses green light. Only one sensor remained: how could the blue light be detected? A few different sensors were considered, but the decision was made to use a **blue** sensor called YF1 since it seemed to have better dynamic performance than the others. Unfortunately, the team was unable to obtain the physical DNA of the YF1 gene and its response regulator, fixJ, because of issues with a material transfer agreement. Instead, the sequences were reconstructed based on the publication and gene synthesis by GenScript, Inc.

Now, the team needed a good color output system. After looking into what was available from the Registry of Standard Biological Parts, it became clear that the only available color reporters were the ones constructed by Cambridge's 2009 iGEM team in their "E. chromi" project (see Lab section 4 of Chapter 5). While this project did provide a way to color bacteria, the pigments were all based on small molecules and required several different genes for the metabolic pathways to produce them. This would make a multichromatic system very large and complex. The ideal solution would be a one-gene–one-color system. The solution came in an article published in the same year where coral chromoproteins had been expressed to color two different strains of bacteria. One of these chromoproteins, amilCP, was **pur-ple-blue**. The other, amilGFP, was **bright yellow** and could work as the output of the green sensor. The bacterial strains were obtained from the Miller lab, and the chromoprotein genes were amplified by PCR, cloned into standard BioBrick™ plasmids, and finally mutagenized to remove a few illegal restriction sites. Expression of these two new genes gave exactly the kind of color output that the team needed. That meant that two colors were ready to use, blue and yellow, while the third color, **red**, already was available in the iGEM distribution kit in the form of mRFP1.

The final system was designed to be modular, so it would be possible to build and test each input and output module separately. Though assembly of the six modules required some 20 BioBrick™ assemblies, all were completed successfully. However, as soon as the modules were assembled, the team discovered several problems with the system. One was bad dynamics of the green sensor: the output promoter PcpcG2 was leaky and gave rather high output signals even when the green sensor was not activated. Also, the red sensor was problematic: it used a native *E. coli* system for regulation, so before it could be tested, the native gene had to be knocked out. Even the promising blue light sensor YF1 had some issues: it showed a rather high fitness cost when expressed, making the bacteria grow very slowly. Still, using a system that linked the blue light input module to a **blue fluorescent** protein output module, it was possible to show that illumination with blue light did indeed activate the output module.

After presenting the project to the iGEM judges during the European Jamboree in Amsterdam in October, the team was selected as one of the finalists that went to the world championships at MIT.

Late during the project, the team garnered a new sponsor, the Korean gene synthesis company Bioneer, and based on the successful use of the amilCP and amilGFP chromoproteins, the team decided to synthesize a few more promising chromoprotein genes for future use (see test tubes on the back cover).

Ultimately, we published all of the chromoproteins and overcame their toxicity problems through engineering (Liljeruhm *et al.*, 2018; Bao *et al.*, 2020). Meanwhile, the Voigt lab created the desired *E. coli* photographic film using non-chromoprotein pigments (Fernandez-Rodriguez *et al.*, 2017).

Uppsala iGEM 2012 — Resistance is Futile

http://2012.igem.org/Team:Uppsala_University

This team chose the medical track, focusing on antibiotic resistance. Would it be possible to construct a system using synthetic small regulatory RNAs (sRNAs; Chapter 2) that could shut down the expression of resistance genes and use this to make antibiotic-resistant bacteria sensitive once again? If such a system could be spread through a bacterial population using a bacteriophage, could it be used as an adjuvant to antibiotics?

Building on an approach previously described, the team linked the 5' untranslated region and the first few codons of different clinical resistance genes to a reporter gene, SYFP2 (Kremers *et al.*, 2006; top right of Fig. 50).

This enabled screening for sRNAs that could down-regulate the expression of the gene. Large random sRNA libraries were generated by BioBrick™ cloning of the native *E. coli* sRNA, spot42 (BBa_K864440, Table 1 in Appendices), and then randomizing its mRNA-binding region by mutagenic PCR (top left of Fig. 50; see Fig. 1b right of Sharma *et al.*, 2011, and also Figs. 11A and 35). These random libraries were transformed into bacteria carrying the reporter gene (SYFP2 translationally fused to the resistance gene), and bacteria showing down-regulation of SYFP2 were sorted out using fluorescence-activated cell sorting (FACS). These cells were then plated on selective agar and visually inspected on a Visi-Blue transilluminator table. Using 460–470 nm blue light, decreased fluorescence of the down-regulated SYFP expression was seen for many colonies! Plasmids encoding sRNAs were prepared from the isolated clones and re-transformed into the strain carrying the reporter gene to verify that the observed lowered fluorescence was actually due to sRNA down-regulation and not due to loss-of-function mutations in the reporter gene. Those sRNAs that still showed efficient down-regulation of expression were sent for sequencing and, based on the results, it was possible to predict the sRNA-mRNA base-pairing interactions through computer modeling. Successful clones were finally transformed into bacteria carrying actual clinical resistance plasmids, and their antibiotic resistances were measured. The results were promising, with some of the sRNAs lowering the resistance of the bacteria by more than 90%. Also, as a side project, the team further expanded their BioBrick™ chromoprotein collection.

Ultimately, we adapted this sRNA approach to targeting chromoproteins in the synthetic biology course (Fig. 35) and generalized the tool further (Vogel *et al.*, 2020).

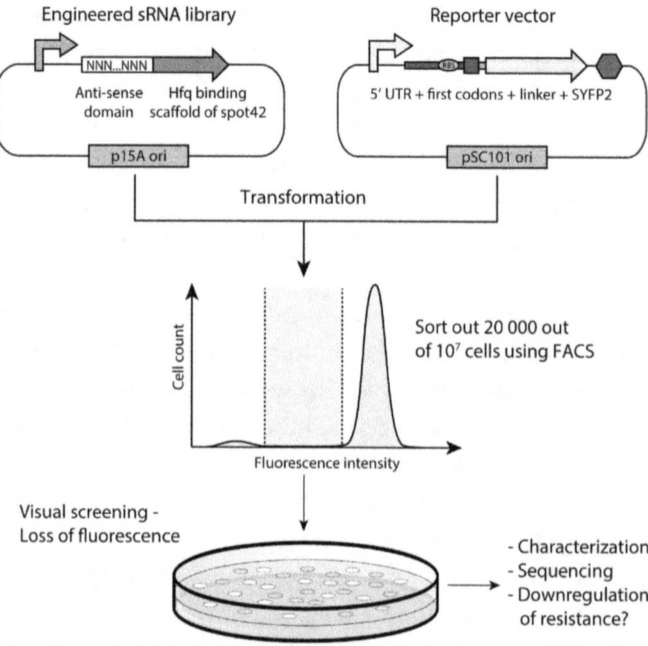

Fig. 50   Screening sRNAs by FACS. *E. coli* was co-transformed with plasmids encoding an sRNA library (top left) and a fluorescent reporter protein fused to the first few codons of the target protein of interest (top right). Only cells with an intermediate fluorescence intensity ($<0.5\%$ of cells) were selected by FACS (middle) for plating (bottom). Later antisense plasmids used a high copy origin (Fig. 35).

9

Appendices

Table 1 BioBrick™ Plasmids Used in the Lab Course. BB codes of parts are those used in the Registry of Standard Biological Parts and link to useful details (e.g. sequences for de novo synthesis, as the Registry no longer distributes many of them). We made available the ideal plasmids for the course in **Addgene's Chromoprotein kit 2.0** (bottom of Table). Most chromoproteins have multiple names, so for clarity the first name given is the original one taken from the reference listed in the same row that reported its amino acid sequence.

Plasmids			Property	Resistance
1. Promoter (including RFP reporter)				
BBa_J23101			High transcription	amp
BBa_J23110			Medium-high transcription	amp
BBa_J23106			Medium transcription	amp
2. Chromoprotein gene (including RBS, not promoter) and reference				
BBa_K1033927	asFP595 (asCP, asPink)*	Lukyanov *et al.* (2000)	Pink-purple color	chloramp
BBa_K1033929	aeCP597 (aeBlue)*	Shkrob et al. (2005)	Blue color	chloramp
BBa_K1033930	amilCP	Alieva *et al.* (2008)	Purple-blue color	chloramp
BBa_K1033931	amilGFP	Alieva *et al.* (2008)	Yellow color	chloramp
BBa_K1033915	amFP486 (amajLime)*	Matz *et al.* (1999)	Lime green color	chloramp
3. Vector backbone (BBa_J04450 RFP reporter)				
pSB1A3			High copy	amp
pSB3A5			Medium-low copy	amp
pSB1C3			High copy	chloramp
pSB3C5			Medium-low copy	chloramp
pSB1K3			High copy	kan
pSB3K3			Medium-low copy	kan

4. Antisense sRNA gene (including promoter)

BBa_K864440	spot42 sRNA	Antisense high copy	chloramp

5. Fluorescent protein genes (if flow cytometer available)

BBa_K864100	Super Yellow Fluorescent Protein 2 (SYFP2)	Yellow fluorescence	chloramp
BBa_K592100	Blue Fluorescent Protein (mTagBFP)	Blue fluorescence	chloramp

Notes: *DNA sequence was codon optimized for *E. coli* commercially; the DNA sequence can be found by accessing the registry (using the BB code) or in Liljeruhm *et al.* (2018).

The Chromoprotein kit 2.0 contains 9 plasmids, sent as bacterial stabs of the *E. coli* strain DH5alpha, found in the four categories below. Please refer to the individual plasmid pages below for more details on each plasmid in this kit:

Promoter (including mRFP1E reporter and Ampicillin resistance)

- promoter J23110 (upstream of RBS, mRFP1)

Chromoprotein gene (including RBS and Chlorampenicol resistance, not promoter)

- asPink chromoprotein promoter-less
- promoter-less mRFP1_Violet chromoprotein
- promoter-less mRFP1_Orange chromoprotein

Chromoprotein expression plasmids (including promoter, RBS, and Chloramphenicol resistance in pSB1C3 vector)

- amajLime chromoprotein
- aeBlue chromoprotein
- amilCP chromoprotein

Vector backbone (including mRFP1E reporter and Kanamycin resistance)

- pSB1K3 containing mRFP1
- pSB3K3 containing mRFP1

Chromoprotein kit 2.0 bacterial stabs.

Tables 2–6   Lists of Essential Equipment and Supplies for 27 Students over 5 Weeks (Protocols 1–11). Key manufacturers are suggested for some items, but there are usually alternatives.

Table 2

Major Equipment	Amount
Agarose gel electrophoresis equipment	6
Bunsen burners and gas packs	8
Centrifuge for 15 mL and 50 mL tubes	1
Heating blocks	4
Ice machine	1
Magnetic stirrer	6
Microcentrifuges for 1.5 mL tubes	4
Micropipettes, 2, 20, 200, 1000 µL	10 each
Microwave oven	1
Oven/incubator (for 42°C)	1
PCR machines	7
pH meter	1
Pipetboys, battery operated, 1–20 mL	8
Scales for centrifuge tubes	1
Shakers and incubator shakers for flasks	3
Spectrophotometer, for 1 mL cuvettes	4
Spectrophotometer, Nanodrop™, for µL samples	1
Vortexer	8
Weighing balances	2

Notes:

Table 3

Minor Equipment	Amount
Beakers, plastic	16
Glasses, orange, Invitrogen Safe Imager	2
Goggles, safety	10
Inoculation loops	8
Inoculation spreaders, glass or metal	8
Lab coats, white	27

Table 3   (*Continued*)

Minor Equipment	Amount
Magnetic stir bars	12
Magnetic stir bar retriever	1
Microcentrifuge rotor adaptors for PCR tubes	20
Notebooks, 80 pages	30
Racks, 0.2, 15, 50 mL tubes	10 each
Racks, 1.5 mL tubes	16
RGB-lamp, LED, Biltema	1
Scissors	4
Spray bottles, plastic, for 75% ethanol	6
Thermometers, alcohol	3
Timers	8

Notes:

Table 4

Disposables	Amount
Bags, plastic	1 box
Cuvettes, plastic, 1 mL	3 boxes
Freezer boxes	16
Gel Extraction Kit, GeneJET (250 preps)	1
Gloves, nitrile, S, M, L	5 each
Gloves, vinyl, S, M, L	15 each
Hazardous waste boxes	5
Marker pens, black	16
Matches	8
Paper sheets, assorted colors, A4 size	1
Parafilm	2
Petri dishes, plastic (10/bag)	50 bags
pH papers, narrow-range	2
Pipettes, glass Pasteur	1 box
Pipettes, glass, 5 mL	8 boxes

(*Continued*)

Table 4 (*Continued*)

Disposables	Amount
Plasmid Miniprep Kit, GeneJET (250 preps)	1
Tape, clear	5
Tape, colored	5
Test tubes, 1.5 mL (500/bag)	4 bags
Test tubes, 15 mL (50/Falcon™ bag)	10 bags
Test tubes, 50 mL (25/Falcon™ bag)	10 bags
Test tubes, PCR (500/bag)	2 bags
Test tubes, PCR, in strips (120/bag)	1 bag
Tip boxes 10, 200, 1000 μL	20 each

Notes:

Table 5

Chemicals	Amount
Agar	500 g
Agarose	200 g
Ampicillin	5 g
ATP, 100 mM, 250 μL	1
Bacto™ tryptone	500 g
Bacto™ yeast extract	500 g
Boric acid	500 g
$CaCl_2$	100 g
Chloramphenicol	5 g
dNTP, 100 mM, 4×250 μL	1
Ethanol, 95%	2 L
Glycerol	500 mL
HCl, conc.	100 mL
Kanamycin	5 g
KCl	100 g
Na_2EDTA or Na_4EDTA	500 g
NaCl	500 g

Table 5 (Continued)

Chemicals	Amount
NaOH	500 g
Sybr® Safe, Invitrogen, 200 μL	2
Tris base	1 kg
Triton®X-100	100 mL

Notes:

Table 6   Molecular Biologicals. We order the enzymes freshly every year. For the cost-conscious, most of them can be overexpressed from plasmids in the annual distribution kit for iGEM teams.

Molecular Biologicals	Amount	Supplier
DpnI FD, 50 U	2	ThermoScientific™
DreamTaq™ DNA polymerase, 200 U	2	ThermoScientific™
EcoRI-HF, heat-inactivatable, 10000 U	1	N.E. Biolabs™
E. coli DH5α cells		
E. coli MG1655 cells		
ladder (5 × 50 μl)	2	ThermoScientific™
Lysozyme, egg white	1 g	Sigma-Aldrich®
Oligo VF2 5′- TGCCACCTGACGTCTAAGAA	25 nmol	Sigma-Aldrich®
Oligo VR 5′- ATTACCGCCTTTGAGTGAGC	25 nmol	Sigma-Aldrich®
Phusion® HF DNA polymerase, 100 U	1	ThermoScientific™
PstI-HF, heat-inactivatable, 10000 U	1	N.E. Biolabs™
SpeI-HF, heat-inactivatable, 500 U	2	N.E. Biolabs™
T4 DNA ligase, 1000 U	1	ThermoScientific™
T4 polynucleotide kinase, 500 U	3	ThermoScientific™
XbaI, heat-inactivatable, 3000 U	1	N.E. Biolabs™

Notes:

Table 7   Schedule for Five-Week Lab Course. Most days began with a 50-minute lecture or tutorial. This was followed by lab work for the rest of the day.

Week 1	Monday	Tuesday	Wednesday	Thursday	Friday
9.15	Lecture	Lecture	Lecture	Tutorial	Lecture
10.15	Lab introduction and safety round	Make agar plates	Tutorial on BioBricks Plasmid preps	Make competent cells	Repeat experiments (if necessary)
12 Lunch					
13.15	Make solutions		BioBrick digestion Analytical agarose gel		
		Practice re-streaking colonies Start overnight cultures	Ligations (if time allows)	Transformations	
~17:00				Plating	Re-streak colonies **Saturday:** Teacher moves plates to 4°C

Week 2	Monday	Tuesday	Wednesday	Thursday	Friday
9.15	Tutorial	Lecture	Lecture	Tutorial	Quiz
10.15	Prepare agarose gel Project designs	Plasmid preps	Tutorial on primer design. Design primers for PCR mutagenesis	Colorimetric assays with lysozyme lysis Read sequences (arrive today or tomorrow)	Colorimetric assays with lysozyme lysis
12 Lunch		Analytical digestion and agarose gel			Read sequences
13.15		1st deadline to send for sequencing.	Order primers	Repeat experiments (if necessary)	
~17:00	Start overnight cultures	Bring computers on next day	Start overnight cultures		

Week 3	Monday	Tuesday	Wednesday	Thursday	Friday
9.15	Lecture	Lecture	Tutorial	Tutorial	Lecture
10.15	Receive mutagenesis primers today	Analytical gel	Ligations	Repeat experiments (if necessary)	Colony PCR
12 Lunch					
13.15	Phosphorylate primers	Dpnl treatment O/N 2nd deadline to send for sequencing	Transformations Plating	Order primers	Analytical gel
~17:00	Start mutagenic PCR			Re-streaking	

Week 4	Monday	Tuesday	Wednesday	Thursday	Friday
9.15	Lecture	Lecture	Tutorial	Tutorial	Lecture
10.15	Mutagenesis	Colorimetric assays	Mutagenesis	Mutagenesis	Mutagenesis
12. Lunch					
13.15		3rd deadline to send for sequencing			
~17.00	Start overnight cultures				

Week 5	Monday	Tuesday	Wednesday	Thursday	Friday
9.15	Tutorial	Lecture	Tutorial	Tutorial	Lecture
10.15	Mutagenesis	Mutagenesis	Write up lab books Prepare lab presentations	Prepare lab presentations	Lab presentations
12. Lunch					
13.15		Final deadline to send for sequencing			
~17.00					

Week 6	Monday	Tuesday	Wednesday		
9.15	Tutorial	Revision tutorial	Revision		
10.15	Complete lab books				
12. Lunch					
13.15			**FINAL EXAM**		
~17.00			End of course		

References

References Cited in Text

Alieva NO, Konzen KA, Field SF, *et al.* (2008) Diversity and evolution of coral fluorescent proteins. *PLoS One* **3**:e2680, 1–12.

Ausubel FM, Brent R, Kingston RE, *et al.* (1999) *Short Protocols in Molecular Biology*, 4th edn. John Wiley & Sons Inc, Hoboken, NJ.

Bao L, Menon PNK, Liljeruhm J, Forster AC. (2020) Overcoming chromoprotein limitations by engineering a red fluorescent protein. *Anal Biochem* **611**:113936, 1–8.

Chaudhari VR, Hanson MR. (2021) GoldBricks: an improved cloning strategy that combines features of Golden Gate and BioBricks for better efficiency and usability. *Synth Biol (Oxf)* **6**:ysab032.

Datta S, Costantino N, Court DL. (2006) A set of recombineering plasmids for gram-negative bacteria. *Gene* **379**: 109–115.

Dedecker P, De Schryver FC, Hofkens J. (2013) Fluorescent proteins: Shine on, you crazy diamond. *J Am Chem Soc* **135**: 2387–2402.

Doudna J, Mali P, eds. (2016) *CRISPR-Cas: A Laboratory Manual*. Cold Spring Harbor Laboratory Press, NY, 192 pp.

Endy D. (2005) Foundations for engineering biology. *Nature* **438**: 449–453.

Fernandez-Rodriguez J, Moser F, Song M, Voigt CA. (2017) Engineering RGB color vision into *Escherichia coli*. *Nat Chem Biol* **13**: 706–708.

Forster AC. (2012) Synthetic biology challenges long-held hypotheses in translation, codon bias and transcription. *Biotechnol J* **7**: 835–845.

Forster AC, Church GM. (2007) Synthetic biology projects in vitro. *Genome Res* **17**: 1–6.

Forster AC, Weissbach H, Blacklow SC. (2001) A simplified reconstitution of mRNA-directed peptide synthesis: activity of the epsilon enhancer and an unnatural amino acid. *Anal Biochem* **297**: 60–70.

Freemont PS, Kitney RI, eds. (2012) *Synthetic Biology: A Primer*. Imperial College Press, London.

Gardner TS, Cantor CR, Collins JJ. (2000) Construction of a genetic toggle switch in Escherichia coli. *Nature* **403**: 339–342.

Gibson DG, Young L, Chuang RY, *et al.* (2009) Enzymatic assembly of DNA molecules up to several hundred kilobases. *Nat Methods* **6**: 343–345.

Goodman DB, Church GM, Kosuri S. (2013) Causes and effects of N-terminal codon bias in bacterial genes. *Science* **342**: 475–479.

Hemsley A, Arnheim N, Toney MD, *et al.* (1989) A simple method for site-directed mutagenesis using the polymerase chain reaction. *Nucleic Acids Res* **17**: 6545–6551.

Hoops S, Sahle S, Gauges R, *et al.* (2006) COPASI: a COmplex PAthway SImulator. *Bioinformatics* **22**: 3067–74.

Hsueh JC, Helmke KJ. (2014) Book review of Synthetic Biology: A Lab Manual. ACS *Synth Biol* **3**: 939–940.

Jinek M, Chylinski K, Fonfara I, *et al.* (2012) A programmable dual-RNA-guided DNA endonuclease in adaptive bacterial immunity. *Science* **337**: 816–821.

Knight TF. (2003) Idempotent vector design for standard assembly of BioBricks. Available at http://hdl.handle.net/1721.1/21168.

Kremers G-J, Goedhart J, van Munster EB, Gadella TWJ. (2006) Cyan and yellow super fluorescent proteins with improved brightness, protein folding, and FRET Förster radius. *Biochem* **45**: 6570–6580.

Kuldell N, Bernstein R, Ingram K, Hart KM. (2015) *BioBuilder*. O'Reilly Media, Sebastopol, USA; 223 pp.

Liljeruhm J, Funk SK, Tietscher S, *et al.* (2018) Engineering a palette of eukaryotic chromoproteins for bacterial synthetic biology. *J Biol Eng* **12**:8, 1–10.

Liljeruhm J, Gullberg E, Forster AC. (2014) *Synthetic biology: A Lab Manual*. World Scientific Press, Singapore; 204 pp.

Lukyanov KA, Fradkov AF, Gurskaya NG, *et al.* (2000) Natural animal coloration can be determined by a nonfluorescent green fluorescent protein homolog. *J Biol Chem* **275**: 25879–25882.

Matz MV, Fradkov AF, Labas YA, *et al.* (1999) Fluorescent Proteins from nonbioluminescent *Anthozoa* species. *Nat Biotechnol* **17**: 969–973.

Na D, Yoo SM, Chung H, *et al.* (2013) Metabolic engineering of *Escherichia coli* using synthetic small regulatory RNAs. *Nat Biotechnol* **31**: 170–174.

Oldenburg KR, Vo KT, Michaelis S, Paddon C. (1997) Recombination-mediated PCR-directed plasmid construction in vivo in yeast. *Nucleic Acids Res* **25**: 451–452.

Quillin ML, Anstrom DM, Shu X, *et al.* (2005) Kindling fluorescent protein from *Anemonia sulcata*: dark-state structure at 1.38 Å resolution. *Biochem* **44**: 5774–5787.

Sambrook J, Russel DW. (2006) *The Condensed Protocols: From Molecular Cloning: A Laboratory Manual.* Cold Spring Harbor Laboratory Press, New York, NY.

Sharma V, Yamamura A, Yokobayashi Y. (2011) Engineering artificial small RNAs for conditional gene silencing in *Escherichia coli*. *ACS Synth Biol* **1**: 6–13.

Shepherd TR, Du L, Liljeruhm J, *et al.* (2017) De novo design and synthesis of a 30-cistron translation-factor module. *Nucleic Acids Res* **45**: 10895–10905.

Shetty RP, Endy D, Knight TF. (2008) Engineering BioBrick vectors from BioBrick parts. *J Biol Engin* **2**:5, 1–12.

Shetty RP, Lizarazo M, Rettberg R, Knight TF. (2011) Assembly of BioBrick standard biological parts using three antibiotic assembly. *Meth Enzymol* **498**: 311–326.

Shkrob MA, Yanushevich YG, Chudakov DM, *et al.* (2005) Far-red fluorescent proteins evolved from a blue chromoprotein from *Actinia equina*. *Biochem J* **392**: 649–654.

Subach OM, Gundorov IS, Yoshimura M, *et al.* (2008) Conversion of red fluorescent protein into a bright blue probe. *Chem Biol* **15**: 1116–1124.

Tou CJ, Kleinstiver (2024) Programmable enzymes for large genome edits. *Nature* **630**: 827–828.

Vogel C, Gynnå A, Yuan J, *et al.* (2020) Rationally-designed Spot 42 RNAs with an inhibition/toxicity profile advantageous for engineering *E. coli*. *Engin Rep* **2**:3, e12126, 1–10.

Wang HH, Isaacs FJ, Carr PA, *et al.* (2009) Programming cells by multiplex genome engineering and accelerated evolution. *Nature* **460**: 894–898.

Wäneskog M, Bjerling P. (2014) Multi-fragment site-directed mutagenic overlap extension polymerase chain reaction as a competitive alternative to the enzymatic assembly method. *Anal Biochem* **444**: 32–37.

Websites in Synthetic Biology

BioBricks Foundation: Available at http://www.biobricks.org

BioBuilder: Available at http://www.biobuilder.org

iGEM Competitions; Synthetic Biology Based on Standard Parts: Available at http://www.partsregistry.org/Main_Page , http://igem.org/Start_A_Team

OpenWetWare: Available at http://www.openwetware.org

Registry of Standard Biological Parts: Available at http://parts.igem.org/Main_Page

Synthetic Biology Engineering Research Center: Available at http://www.synberc.org

Synthetic Biology Open Language Visual Standard: Available at http://www.sbolstandard.org/visual

Text Boxes

Index